MRINAL K. DAS

D0793311

CELL FINE STRUCTURE

An Atlas of Drawings of Whole-Cell Structure

THOMAS L. LENTZ, M.D.

Department of Anatomy,
Yale University School of Medicine

W. B. SAUNDERS COMPANY · PHILADELPHIA · LONDON · TORONTO

1971

W. B. Saunders Company: West Washington Square
Philadelphia, Pa. 19105

12 Dyott Street
London, WC1A   1DB

1835 Yonge Street
Toronto 7, Ontario

Cell Fine Structure—*An Atlas of Drawings of Whole-Cell Structure*   SBN 0-7216-5718-4

Print No.:   9   8   7   6   5   4   3   2   1

To Judy, Stephen, Christopher, and Sarah

A set of 35 mm slides of the drawings in this Atlas
is available from the publisher

PREFACE

This atlas is a collection of diagrams summarizing present information on the fine structure of mammalian cell types. In the last ten years especially, improved methods of fixation and embedding of tissues for electron microscopy have permitted definitive characterization of the fine structural organization of cells. Information on biological ultrastructure has accumulated at a rapid rate, and virtually all cell types have been examined with the electron microscope. Not all cells, though, have been described to the same degree of accuracy and thoroughness, and much remains to be accomplished in cell fine-structural morphology. Much of the information on cell fine structure is surveyed in several excellent texts and atlases. However, because of the nature of the procedures necessary to view tissues with the electron microscope, it is difficult to obtain a comprehensive picture of an entire cell and all its components in a single electron micrograph. Thin sections reveal only a tiny fraction of a tissue or even of an individual cell, and only rarely does a single micrograph illustrate to advantage all the characteristic structures and organelles of a cell. Usually it is necessary to examine many sections to form a clear understanding of the nature of a particular cell type. The use of diagrams, however, permits all the information available in a number of electron micrographs to be included in a single illustration. Thus, the diagrams in this atlas represent idealized cell types and include the variations encountered in thin sections of different regions of a tissue, during different functional states or in different species.

This volume contains 184 diagrams representing most of the major cell types of the body. The illustrations are grouped under the following tissue and organ systems: Blood, bone marrow, and lymphatic tissue, Connective tissue, Muscular tissue, Vascular tissue, Skin and other epithelia, Digestive system, Respiratory system, Urinary system, Male reproductive system, Female reproductive system, Embryonic tissues, Endocrine system, Nervous system, Eye, and Ear. Each illustration is accompanied by a description of the cytological structure of the cell. Although cytological structure is emphasized, functional correlations are made where these are pertinent and reasonably well-established. Selected references to major morphological studies are included.

As an inclusive survey of the cytology of all cell types, this volume is intended primarily for students of cell biology at the college or medical school level. Often in making the transition from the histological study of the cell with the light microscope to consideration of the individual organelles and components at the submicroscopic level, the cytological organization of individual cell types in their entirety is neglected. Thus, it is hoped that this atlas will fill this gap to some extent by presenting fine structure from the point of view of the organization of the whole cell. Although every attempt has been made to accurately represent structures as seen with the electron microscope, the diagrams should not be used as a substitute for study of actual micrographs but rather as an introduction or aid in their interpretation. These illustrations, compiled in notebooks, have indeed been used in this manner for the last few years in the Cell Biology course given at the Yale University School of Medicine. This volume should also be of use as a reference to those in other fields who might occasionally require some information on the cytology of a particular cell type with which they are unfamiliar.

New Haven, Connecticut   THOMAS L. LENTZ

CONTENTS

CONTENTS

CONTENTS

CONTENTS

X

CONTENTS

xi

A	A band	Gly	Glycogen
Ac	Acrosome	Gr	Granule
AcG	Acrosomal granule	H	H band
AcV	Acrosomal vesicle	Hb	Hemoglobin
AG	Azurophil granule	I	I band
AL	Annulate lamella	IB	Intercellular bridge
AN	Accessory nucleolus	IF	Inner fiber
An	Annulus	IS	Inner segment
AP	Apical pit	JF	Junctional fold
AS	Alveolar space	Kc	Kinocilium
AV	Apical vesicle	KG	Keratohyalin granule
Ax	Axon	LC	Longitudinal column
BB	Basal body	LD	Lipid droplet
BC	Bile canaliculus	LPG	Lipofuscin pigment granule
BL	Basement lamina	Lu	Lumen
BP	Basal plate	Ly	Lysosome
C	Cilium	M	Mitochondrion
CA	Compound aggregate	M	M line
Ca	Canaliculus or canal	MA	*Macula adherens*
Cap	Capitellum	Ma	Mesaxon
CB	Chromatoid body	Mb	Microbody
CD	Colloid droplet	Mel	Melanosome
Ce	Centriole	MF	Marginal fold
CeS	Centriolar satellite	Mf	Myofilament
CG	Cisternal granule	MG	Mucous granule
Ch	Channel	MO	*Macula occludens*
Chy	Chylomicron	MP	Middle piece
CL	Capillary lumen	MS	Myelin sheath
Co	Collagen	Mt	Microtubule
CP	Cuticular plate	Mv	Microvillus
Cry	Crystal or crystalloid	MvB	Multivesicular body
CV	Condensing vacuole	N	Nucleus
Cy	Cytosome	NCS	Nucleolar channel system
Den	Dendrite	NE	Nuclear envelope
DT	Dendritic thorn	NF	Nerve fiber
DV	Dense core vesicle	Nf	Neurofilament
EC	Extraneous coat	NI	Nuclear inclusion
EF	End foot	NiS	Nissl substance
EP	End piece	Nk	Neck
ER	Endoplasmic reticulum	Nl	Nucleolus
F	Fibril	Nln	Nucleolonema
FA	*Fascia adherens*	NP	Nuclear pore
Fe	Ferritin	NS	Nucleolar satellite
Fen	Fenestra	NsG	Neurosecretory granule
FiS	Filtration slit	ODF	Outer dense fiber
FL	Fibrous lamina	OF	Outer fiber
Fl	Filament	OS	Outer segment
FO	*Fascia occludens*	Ot	Otolith
FP	Foot process	OV	Olfactory vesicle
FS	Fibrous sheath	PA	Pars amorpha
FV	Fusiform vesicle	PaG	Proacrosomal granule
G	Golgi apparatus	PcB	Paracentriolar body

ABBREVIATIONS

Pd	Pseudopodium	SpG	Specific granule
PDC	Platelet demarcation channel	SR	Sarcoplasmic reticulum
PG	Protein granule	SsC	Subsurface cisterna
PhV	Phagocytic vacuole	SsW	Subsynaptic web
PiG	Pigment granule	StC	Striated column
PM	Plasma membrane	SV	Synaptic vesicle
PP	Principal piece	SxV	Sex vesicle
PV	Pinocytotic vesicle	Sy	Synapse
R	Ribosome	Tf	Tonofilament
RC	Ring centriole	TM	Tectorial membrane
RS	Rod spherule	TP	Tongue process
Rt	Rootlet	Tr	Triad
SA	Spine apparatus	TS	Transverse system
SB	Synaptic bar	Tu	Tubule
SC	Schwann cell	TW	Terminal web
Sc	Stereocilium	V	Vesicle
SER	Smooth endoplasmic reticulum	Vac	Vacuole
SG	Secretory granule	Z̲	Z line
SH	Sensory hair	ZA	*Zonula adherens*
Si	Sinusoid	ZG	Zymogen granule
SnC	Synaptonemal complex	ZO	*Zonula occludens*
Sp	Spine		

GENERAL ORGANIZATION
OF THE CELL

GENERAL ORGANIZATION OF THE CELL

Cells occur in a wide assortment of shapes and sizes, differ in their structural and chemical components, and perform a variety of complex functions necessary for the survival of the organism. Despite these obvious differences, cells have certain basic similarities in their general organization. Thus, all cells have a limiting plasma membrane, nearly all have a nucleus, and most contain, in varying amounts and proportions, ribosomes, endoplasmic reticulum, Golgi apparatus, and mitochondria. Other structures, such as lysosomes, microtubules, filaments, centrioles, and inclusions, are frequently but not always present. Although these components are specialized to different degrees in different cell types, each has certain common properties and performs basically similar functions in all cells. The organization of the common constituents of the cell is described here, as an introduction to the study of specialized cell types.

The cell is bounded by a limiting membrane, the plasmalemma or plasma membrane, which has an overall diameter of about 100 Å (Figs. 1, 2, PM). At high magnifications, the membrane appears to be composed of two dense lamellae separated by a clear layer (see Plates 78, 158, 159). These layers constitute the trilaminar unit membrane and are generally thought to be two protein layers and an intervening bimolecular lipid layer.

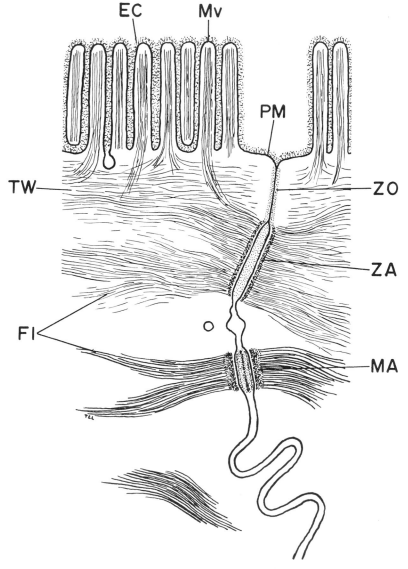

Figure 1. See text for discussion.

GENERAL ORGANIZATION OF THE CELL

The cell surfaces show a variety of specializations related to functions such as absorption, secretion, fluid transport, and adhesion. Microvilli (Figs. 1, 3, Mv) are slender processes found on cells whose principal function is absorption. They may be numerous and closely and regularly packed in some cells (intestinal epithelial cell, Plate 76; proximal convoluted tubule of the kidney, Plate 99), but in a large number of other cells they are scattered over the surface and are of different lengths. Microvilli clearly amplify the absorptive surface area of the cell. Increases in surface area are formed at the opposite pole of the cell by complicated infoldings or processes of the basal surface (Fig. 2). Basal infoldings are commonly found in cells engaged in active transport of fluids and ions (clear cell of sweat gland, Plate 55; striated duct cell of the salivary glands, Plate 68; distal convoluted tubule of the kidney, Plate 101; ciliary epithelial cell, Plate 171; marginal cell of the stria vascularis, Plate 176; dark cell of the crista ampullaris, Plate 182).

Transport of fluids, proteins, and other substances takes place in pinocytotic vesicles (Fig. 3, PV) that form as small invaginations of the surface and pinch off to lie free in the cytoplasm (see Plates 12 and 42). When the surface of the cell has an extraneous coat, it often lines the small invaginations and serves as a site for selective adsorption of substances (see Plate 12). Some vesicles have short, radially arranged bristles extending from the outer leaflet of the unit membrane. These vesicles are known as coated vesicles; when associated with the cell surface (Fig. 3), they are believed to be involved in the selective uptake of substances, especially proteins. After these vesicles pinch off from the surface, they fuse to form larger vacuoles or granules. Similar, coated vesicles

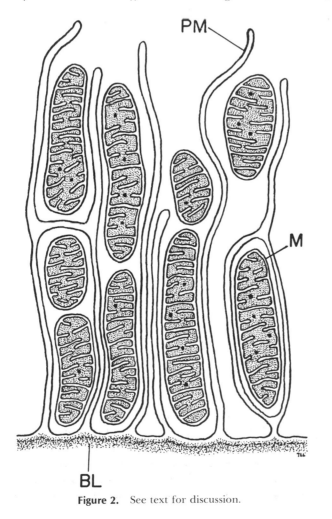

Figure 2. See text for discussion.

3

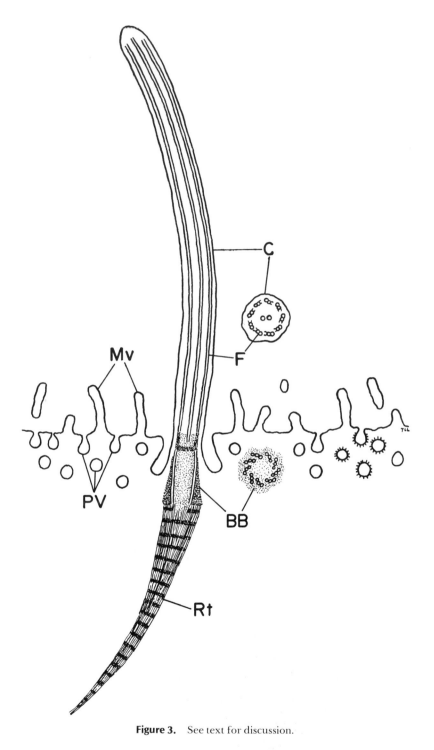

Figure 3. See text for discussion.

often occur in the Golgi region, and these function in the transport of lytic enzymes to lysosomes (Fig. 7).

Junctional complexes are formed along the apical lateral borders of cells comprising continuous linings or epithelia (see Plate 78). The junctional complex (Fig. 1) consists of an apical tight junction or *zonula occludens* (ZO), an intermediate region or *zonula adherens* (ZA), and a desmosome or *macula adherens* (MA). Junctions occurring between other types of cells are often variations of one or more components of the basic tripartite junctional complex. Another type of junction is the close or gap junction.

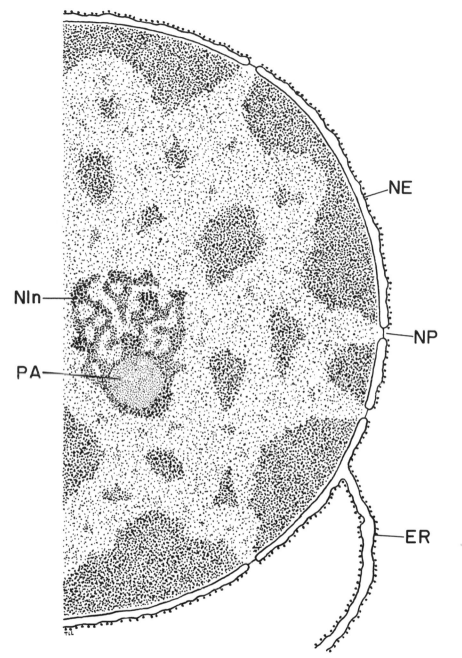

Figure 4. See text for discussion.

Here, the adjacent plasma membranes are very closely apposed but remain separated by a thin intercellular space of 20 to 30 Å. Materials may be associated with the plasma membrane, such as extraneous coats (Fig. 1, EC) of filamentous material and the basement membrane or lamina (Fig. 2, BL). The latter is a thick coat containing filamentous material that parallels the basal plasma membrane of the cell but is separated from it by a clear space.

Cells also possess motile specializations of their surfaces in the form of cilia (Fig. 3, C) and flagella (olfactory cell, Plate 86; ciliated cell of the respiratory epithelium, Plate 91; spermatid, Plate 106; spermatozoon, Plate 107; ciliated cell of the efferent ductule, Plate 110; ciliated cell of the female reproductive tract, Plate 121; ependymal cell, Plate 165). The cilium is bounded by the plasma membrane and contains nine peripheral

5

pairs of hollow fibrils and a central pair of fibrils (see Plate 107). The fibrils (F) terminate in a basal body (BB) situated in the cytoplasm just below the cell surface. From the basal body, a striated rootlet (Rt) composed of fine filaments extends into the cytoplasm.

The basal body of the cilium (Fig. 3, BB) appears to be identical to the centriole. In cells where the centrioles are not situated near the cell surface, a pair of centrioles (the diplosome) (Fig. 5, Ce) is commonly found near the nucleus and is partially surrounded by the Golgi apparatus. The centrioles are arranged at right angles to one another and are composed of nine triplets of hollow fibrils. Some centrioles have small dense appendages, or centriolar satellites (CeS), from which microtubules (Mt) radiate into the cytoplasm. Centrioles are capable of replicating and, besides giving rise to cilia, may serve as organizers for microtubules.

The largest and most conspicuous component of the cell is the nucleus (Fig. 4). Only a few cells lack nuclei (erythrocyte, Plate 16; platelet, Plate 18; epidermal cell of the stratum corneum, Plate 51; lens epithelial cell, Plate 172), and these cells are end stages in differentiative processes. The nucleus is bounded by the nuclear envelope (NE), which is composed of two parallel membranes enclosing the perinuclear space. The outer membrane, facing the cytoplasm, bears ribosomes and in places is continuous with the cisternae of the rough-surfaced endoplasmic reticulum (ER). In some places, the nuclear envelope is perforated by circular openings or nuclear pores (NP). The inner and outer nuclear membranes are continuous around the circumference of the pores. The pores appear to be bridged by a thin septum or diaphragm. The nucleus contains chromatin and the nucleolus embedded in the nuclear matrix or nucleoplasm. Chromatin is the chromosomal substance of the nucleus and consists of deoxyribonucleic acid (DNA) combined with histones and other proteins. It exists in a highly condensed form (heterochromatin), or it is dispersed (euchromatin). Heterochromatin is more abundant in cells that are relatively inert metabolically (orthochromatic erythroblast, Plate 14; spermatozoon, Plate 107), whereas euchromatin, active in the synthesis of messenger ribonucleic acid (RNA), predominates in metabolically or synthetically active cells (basophilic erythroblast, Plate 12; spermatocyte, Plate 105; neuron, Plate 157). The nucleolus is a rounded body and is involved in the elaboration of ribosomal RNA. It is usually composed of a round, finely granular pars amorpha (PA) and the nucleolonema (Nln). The nucleolonema is composed of filamentous material that may be either compact or arranged in coarse anastomosing strands. Dense granules similar in appearance to cytoplasmic ribosomes are embedded in this filamentous material. The nucleolus is most prominent in dividing or undifferentiated cells (hemocytoblast, Plate 1; spermatocyte, Plate 105; oocyte, Plate 116; embryonic cells, Plates 126 and 127) and in cells active in protein synthesis (basophilic erythroblast, Plate 12; pancreatic acinar cell, Plate 84; endometrial epithelial cell, Plate 122).

The structures and organelles of the cell are embedded in the cytoplasmic matrix or hyaloplasm. Although largely structureless when viewed with the electron microscope, the hyaloplasm is an important constituent of the cell and contains water, ions, and soluble enzymes and proteins. Differences in the density of the hyaloplasm in different cells suggest functional specializations of the matrix.

Ribosomes are small particles of ribonucleoprotein, about 150 Å in diameter. High magnification shows them to be composed of a large and a small subunit. Ribosomes (Fig. 7, R) may occur free and singly in the hyaloplasm, or in clusters known as polyribosomes. Polyribosomes are composed of individual ribosomes joined by a thin strand of messenger RNA and are often arranged in spirals, rosettes, or circles. Free ribosomes are most common in undifferentiated cells (hemocytoblast, Plate 1; lympho-

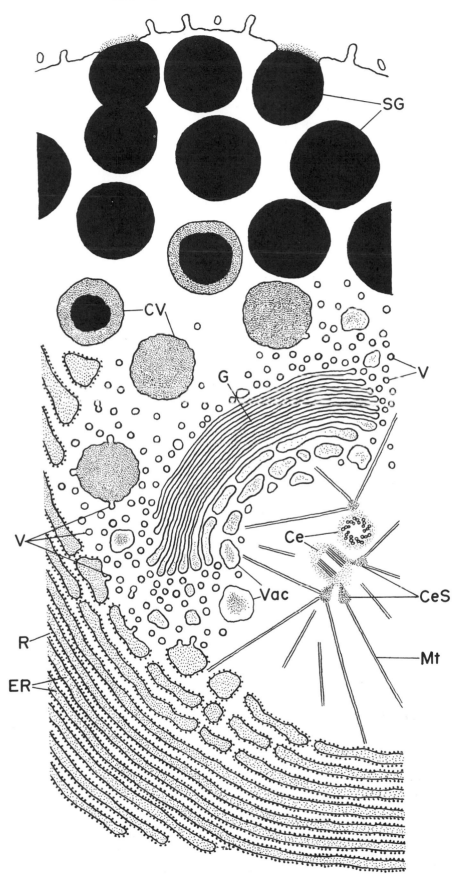

Figure 5. See text for discussion.

GENERAL ORGANIZATION OF THE CELL

cyte, Plate 10; spermatocyte, Plate 105; embryonic cells of of the blastocyst, Plates 126 and 127) and in cells synthesizing protein that is retained within the cell for endogenous use (erythroblast, Plates 12 to 14; stratum germinativum, Plate 48).

The rough-surfaced endoplasmic reticulum (Fig. 5, ER) is formed by membranous sacs, or cisternae, bearing ribosomes (R) on their outer surfaces. In some cells, the cisternae are flattened and stacked in parallel and occupy much of the cytoplasm, especially in the basal regions of the cell. Whereas free ribosomes are commonly associated with endogenous protein synthesis, rough-surfaced endoplasmic reticulum is most extensive in cells that export the protein synthetic product outside the cell. Polypeptides are synthesized on the ribosomes and transferred through the endoplasmic reticulum membrane into the cisternae. Particulate material representing newly formed proteins can sometimes be seen within the cisternae.

In some cells, the endoplasmic reticulum lacks ribosomes and is known as smooth-surfaced endoplasmic reticulum (Fig. 6, SER). The agranular membranes are usually in the form of branching and anastomosing tubules. The tubules may be closely packed to form regular arrangements or whorls, and they are sometimes arranged around mitochondria and lipid droplets. In places, the smooth tubules are continuous with rough cisternae (ER). Rough endoplasmic reticulum is most prominent in protein-synthesizing cells, but smooth endoplasmic reticulum is commonly found in steroid-synthesizing cells (interstitial cell of the testis, Plate 109; lutein cell, Plate 119; cells of the adrenal cortex, Plates 146 to 148) and in cells engaged in lipid metabolism and synthesis (brown adipose cell, Plate 29; sebaceous gland cell, Plate 57; intestinal epithelial cell, Plate 77; hepatic cell, Plate 81).

The Golgi apparatus (Fig. 5, G) is usually situated between the nucleus and the apex of the cell. It is composed of a stack of parallel, smooth membranous lamellae, small vesicles (V), and vacuoles (Vac). The vesicles are concentrated at the ends of the lamellae and along the outer or convex surface of the Golgi. Vacuoles are larger and often arranged along the inner or concave face of the Golgi. Both vesicles and vacuoles are sometimes continuous with the ends of the lamellae. The Golgi apparatus is the site of packaging and further concentration of the proteins synthesized in the rough-surfaced endoplasmic reticulum. The proteins reach the Golgi within small vesicles (V) that have pinched off from rough-surfaced lamellae adjacent to the Golgi. Fusion of vesicles with condensing vacuoles (CV) concentrates the synthetic product. Further concentration of secretory product gives rise to mature secretory granules (SG), which are stored in the apex of the cell until discharge occurs by fusion of the membrane enclosing the secretory granule with the plasma membrane. Temporary storage of secretory products is characteristic of many cells (sweat gland cells, Plates 54 and 56; salivary gland cells, Plates 65 and 66; chief cell, Plate 72; Paneth cell, Plate 79; goblet cell, Plate 80; other mucous cells, Plates 70, 71, 75, 92; pancreatic acinar cell, Plate 84; seminal vesicle secretory cell, Plate 113; prostatic epithelial cell, Plate 115; secretory cell of the oviduct, Plate 120; and many endocrine cells, Plates 133 to 140, 142 to 144, 149 to 153). Other cells synthesizing and exporting proteins do not store the products and are therefore largely devoid of condensing vacuoles and secretory granules (plasma cell, Plate 19; fibroblast, Plate 25; chondrocyte, Plate 30). Materials synthesized by these latter cells appear to be released as soon as they are formed; the Golgi vesicles or vacuoles migrate to the cell surface, fuse with the plasma membrane, and release their contents (reverse pinocytosis).

Besides serving to concentrate proteins, the Golgi apparatus may play a role in the synthesis of mucopolysaccharides (polysaccharides containing hexosamine), and mucoproteins and glycoproteins (hexosamine-containing polysaccharides in firm chemical union with a peptide). Precursors (e.g., glucose) of the polysaccharides are incorporated by the Golgi complex, and synthesis of the carbohydrate moieties presumably

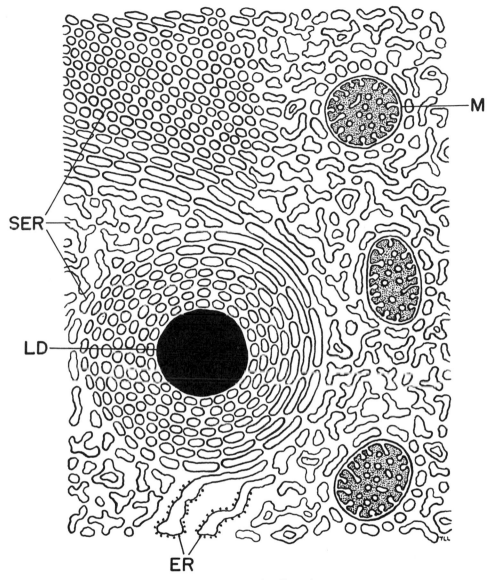

Figure 6. See text for discussion.

takes place at this site. The protein components are synthesized by the ribosomes of the rough-surfaced endoplasmic reticulum and then are combined with the polysaccharides in the Golgi. Cells in which this mechanism takes place include the various types of mucus-secreting cells, plasma cell (Plate 19), chondrocyte (Plate 30), cells of the anterior pituitary that produce the glycoprotein hormones FSH, LH, and TSH (Plates 135 to 137), and the follicular cell of the thyroid (Plate 142). Sulfation of mucopolysaccharides also occurs in the Golgi complex (see chondrocyte, Plate 30).

Mitochondria vary in shape from spherical to elongated rods (Figs. 2, 6, and 7, M). They are bounded by two membranes: a continuous outer and inner mitochondrial membrane. The inner mitochondrial membrane is thrown up into folds called cristae, which project into the central matrix cavity of the mitochondrion. The matrix may be composed of moderately dense material and usually contains opaque granules that may bind divalent cations. A major function of mitochondria is generation of energy in the form of adenosine triphosphate (ATP) by oxidative phosphorylation. The soluble enzymes of the Krebs and fatty acid cycles reside in the matrix, whereas the enzymes of the respiratory pathway (electron transport chain) are located on the inner membrane.

GENERAL ORGANIZATION OF THE CELL

Both DNA and RNA have been identified in the matrix of mitochondria. Mitochondria are most abundant in cells with high energy requirements (brown adipose cell, Plate 29; cardiac muscle fiber, Plate 38). In cells engaged in active transport, the mitochondria are frequently aligned parallel to the membranes within the basal processes (Fig. 2), where they provide a ready source of energy. The cristae of mitochondria are usually lamellar in form, but in some cells engaged in steroid synthesis (lutein cell, Plate 119; cell of the zona fasciculata, Plate 147) they are tubular or vesicular (Fig. 6).

Lysosomes are membrane-limited dense granules that contain a number of hydrolytic enzymes (Fig. 7, Ly). Some of the enzymes identified in lysosomes are acid phosphatase, ribonuclease, deoxyribonuclease, cathepsin, collagenase, glucosidase, glucuronidase, and aryl-sulphatase. Lysosomes function in the breakdown and digestion of materials taken into the cell by phagocytosis (see polymorphonuclear leukocyte, Plate 5) or in the disposal of cell organelles or secretory products. Lysosomes may be released outside the cell, effecting digestion of extracellular materials (see osteoclast, Plate 32). The class of bodies containing hydrolytic enzymes is quite heterogeneous in fine structure, which depends largely on the stage of the digestive process. The enzymes comprising lysosomes are considered to be synthesized on ribosomes of the endoplasmic reticulum, transferred to the Golgi apparatus, and packaged into granules having a finely granular matrix. It is these structures that are usually called lysosomes or, often, dense bodies. Golgi vesicles, often coated, that contain enzymes can contribute to the lysosomes or form membrane-limited aggregates or multivesicular bodies (Fig. 7, MvB). Particles are incorporated into the cell within phagocytic vacuoles. Lysosomes fuse with these vacuoles and form digestive vacuoles where breakdown of the ingested material occurs. Cell organelles are segregated in autophagic vacuoles to which lysosomes also fuse. The end product of both processes is a residual body containing indigestible remnants. The residual body often occurs as a lipofuscin pigment granule (Fig. 7, LPG), which in certain cells accumulate with age (cell of the zona reticularis of the adrenal cortex, Plate 148; neuron, Plate 157).

Microtubules (Fig. 5, Mt) are straight, hollow cylinders with a diameter of about 200 Å. Their wall is composed of 13 filamentous subunits. They are especially common in cells with asymmetric processes (olfactory cell, Plate 86; glomerular epithelial cell, Plate 98; spermatid, Plate 106; neuron and its processes, Plates 157 to 160; lens epithelial cell, Plate 172; supporting cells, Plates 175 and 181) and are thought to play a role in maintenance of cell shape (see erythroblast, Plates 12 to 14; and platelet, Plate 18). Microtubules have also been associated with cytoplasmic streaming and mediation of the movement of vesicles and other particles. In dividing cells, they form the mitotic spindle and may be responsible for the movement of the chromosomes. They are common in the centrosphere region of the cell, where they terminate on the centriolar satellites. Microtubules seem to be identical to the fibrils of motile cilia and flagella.

Many cells contain fine filaments with a diameter of 40 to 100 Å (Fig. 1, Fl). These course in the cytoplasm between other organelles and are often organized into bundles or fibrils. They are especially common in cells of the epidermis (Plates 48 to 51) and in other epithelial cells. They are often associated with the *zonula adherens* and *macula adherens* of the junctional complex; and, in the apex of the cell, they form a terminal web (Fig. 1, TW) from which other organelles are largely excluded. In neurons, they extend into the processes as neurofilaments (Plates 157 to 160), and some astrocytes are especially rich in filaments (Plates 161 and 162). Special types of filaments, thick myosin filaments and thin actin filaments, form the contractile system of muscle (see Plates 34, 35, 38, 40, 41).

Cell inclusions are considered to be inactive, storage forms of secretory products or metabolites, as opposed to the organelles, which are active in the metabolism of the cell.

10

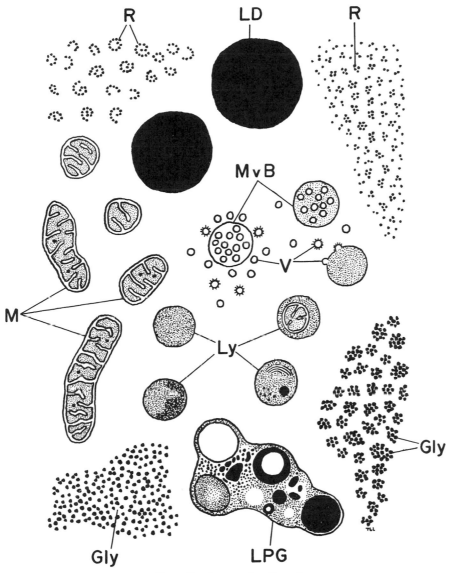

Figure 7. See text for discussion.

Although it is often difficult to make a distinction between organelles and inclusions, examples of the latter are secretory granules, lipid droplets, glycogen granules, pigment granules, and crystals. Lipid is usually stored in the form of large spherical droplets (Figs. 6, 7, LD); (see also white adipose cell, Plate 28; red muscle fiber, Plate 35; interstitial cell, Plate 109; lutein cell, Plate 119; adrenal cortical cells, Plates 146 to 148). Glycogen occurs as dense granules about 300 Å in diameter (Fig. 7, Gly); (see also hepatocyte, Plate 81; decidual cell, Plate 124). The granules may be dispersed (beta particles), or aggregated into rosettes (alpha particles).

The cells of the body differ in the type, number, and organization of the basic organelles and inclusions. Most highly specialized structures, such as annulate lamellae, cytosomes, microbodies, myelin, stereocilia, sensory hairs, and subsurface cisternae, to name a few, are modifications of the simpler organelles and will be described as they are encountered in particular cells. The characteristic differences in specialization and pattern of organization are the basis for the identification of the cell types in the following plates. Differences in the functions carried out by the cells, furthermore, are largely determined by the relative number and distribution of the common constituents as well as by their degree of structural and enzymatic specialization.

11

REFERENCES

Bloom, W. and D. W. Fawcett, 1968. A Textbook of Histology, 9th ed. W. B. Saunders Company, Philadelphia, 858 pp.

Fawcett, D. W., 1966. The Cell: Its Organelles and Inclusions. W. B. Saunders Company, Philadelphia, 448 pp.

Greep, R. O. (ed.), 1966. Histology, 2nd ed. McGraw-Hill, New York, 914 pp.

Porter, K. R. and M. A. Bonneville, 1968. Fine Structure of Cells and Tissues, 3rd ed., Lea and Febiger, Philadelphia, 196 pp.

Rhodin, J. A. G., 1963. An Atlas of Ultrastructure. W. B. Saunders Company, Philadelphia, 222 pp.

BLOOD, BONE MARROW, AND LYMPHATIC TISSUE

BLOOD, BONE MARROW, AND LYMPHATIC TISSUE

1—HEMOCYTOBLAST

The hemocytoblast or stem cell is an undifferentiated cell of the bone marrow and is capable of giving rise to other cell types. The cell is large (~20 μ) with a spherical or oval nucleus. The nuclear envelope is perforated by many pores. The chromatin is relatively evenly distributed in the nucleoplasm, although some small clumps occur throughout the nucleoplasm and adjacent to the inner aspect of the nuclear envelope. One or more large nucleoli (Nl) are present and consist of filamentous material and granules. The cytoplasm contains a large number of free ribosomes (R). These occur singly or in clusters called polyribosomes. A few elongated cisternae of endoplasmic reticulum that bear ribosomes are also present. Small mitochondria are distributed throughout the cytoplasm and may cluster in the pole of the cell opposite the Golgi apparatus. The Golgi apparatus (G) is situated adjacent to the nucleus and consists of a few membranous lamellae and small vesicles. Large granules characteristic of mature cells do not occur in stem cells. A few small, dense, membrane-bounded granules of unknown nature sometimes occur near the Golgi apparatus. The cavities of the Golgi elements are generally of low density. A pair of centrioles (Ce) is associated with the Golgi apparatus. Microtubules are abundant in this region and also occur elsewhere in the cytoplasm.

The structure of the hemocytoblast is characteristic of undifferentiated cells (lymphocyte, Plate 10; spermatocyte, Plate 105; oocyte, Plate 116; trophoblast cell, Plate 126; inner cell mass cell, Plate 127), which have an abundance of free ribosomes and few membranous elements. It is generally agreed that the myeloid elements of the bone marrow differentiate from the free stem cell referred to as a hemocytoblast; but it is uncertain whether or not this stem cell is identical to the stem cell (lymphoblast) that gives rise to the lymphoid elements. Lymphocytes (Plate 10), particularly large lymphocytes, closely resemble hemocytoblasts, and may be equivalent to them.

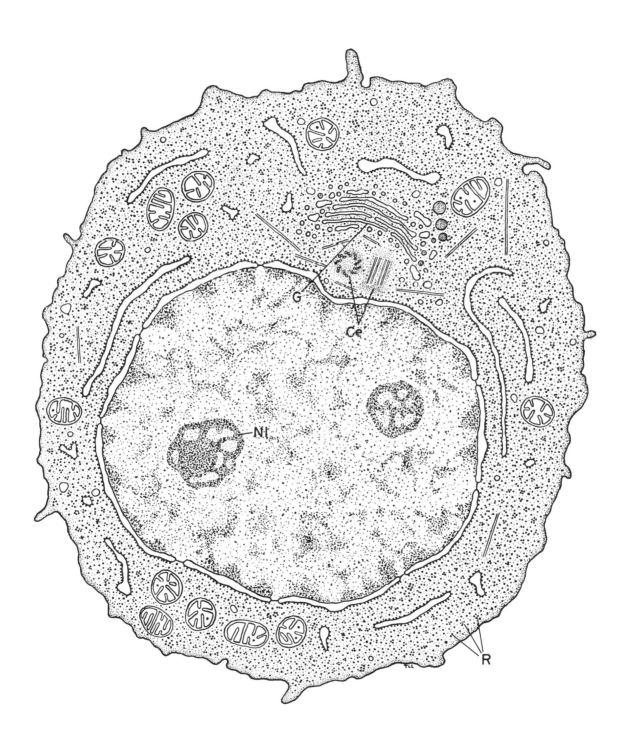

2 — PROGRANULOCYTE

Progranulocytes or promyelocytes are maturing granulocytes that contain azurophil granules (AG). The first specific granules may appear at the progranulocyte stage, but they are few in number. The progranulocyte may reach 18 μ in diameter. It contains a slightly compressed nucleus. The chromatin material is distributed throughout the nucleoplasm in many small clumps and shows some condensation on the inner aspect of the nuclear envelope. Although still prominent, the nucleolus is not as well defined as in the hemocytoblast. It is composed of a coarse intertwining strand that contains dense granules (nucleolonema, Nln) and surrounds a spherical, finely granular body (pars amorpha, PA). Single, elongated cisternae of rough-surfaced endoplasmic reticulum are abundant in the cytoplasm. In addition, a large number of free ribosomes occur in the cytoplasm, many forming polysomal clusters. Mitochondria are found among the other structures.

A large centrosphere region is situated adjacent to the nucleus. A pair of centrioles occurs in the center of this zone and is surrounded by elements of the elaborate Golgi apparatus. Microtubules (Mt) are abundant in this region and converge on the centriolar satellites (CeS). Stacks of four to nine membranous lamellae are oriented around the centrioles. The proximal or inner cisternae are more dilated than are the flattened distal or outer cisternae. Small vesicles are especially abundant near the ends of the membranous lamellae. Larger vacuoles (Vac) are most common along the proximal or concave face of the Golgi apparatus. These vacuoles are 200 to 400 mμ in diameter and contain a central core of dense material (100 to 150 mμ). The central core is separated from the membrane of the vacuole by a zone of low density that contains flocculent or reticular material. Dense material may occur within the dilated ends of the inner Golgi lamellae. Larger vacuoles (up to 900 mμ) contain several dense cores and are formed by fusion of the small vacuoles. These large vacuoles are immature azurophil granules. Their contents undergo progressive condensation to form mature azurophil granules (AG). The mature granules are 800 mμ in diameter, composed of uniformly dense material, and bounded by a membrane.

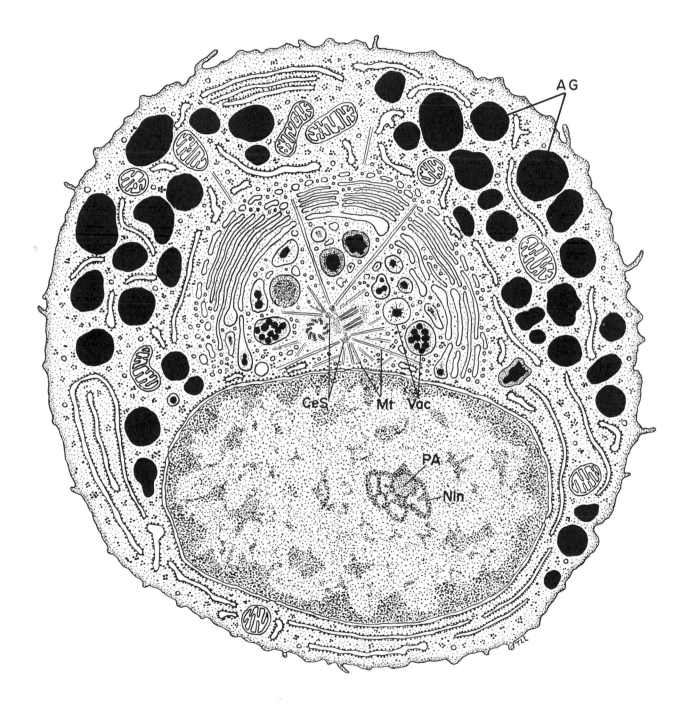

3 — NEUTROPHILIC MYELOCYTE

The myelocyte represents the stage in granulocyte maturation when the specific (neutrophilic, eosinophilic, or basophilic) granules accumulate in greater numbers while the azurophil granules become relatively less abundant. The cell is smaller, 12 to 16 μ in diameter, and contains an irregular or slightly indented nucleus. The chromatin material is condensing and occurs in clumps in the nucleoplasm and adjacent to the nuclear envelope. The nucleolus is inconspicuous or has disappeared. In the cytoplasm, cisternae of rough-surfaced endoplasmic reticulum are present but not as numerous as in the promyelocyte. Free ribosomes are also reduced in number. Mitochondria are not as numerous and are smaller with fewer cristae.

The Golgi zone of myelocytes, although still very extensive, occupies a smaller area and contains fewer lamellae (three to five) in each stack. Centrioles and their associated satellites occur in the centrosphere region. The inner cisternae appear empty and are no longer forming azurophil granules. The outer cisternae of the Golgi apparatus contain finely granular material that becomes dense toward the distal or convex surface. The ends of some of the lamellae are expanded and contain dense material. Small vacuoles containing dense material are situated along the distal face of the Golgi apparatus. These fuse to form larger vacuoles whose contents condense to form mature specific granules (SpG).

Both azurophil (AG) and specific granules are present in myelocytes. The azurophil granules are mature, or nearly so. Immature specific granules, most numerous in the vicinity of the Golgi apparatus, contain partially condensed material and, sometimes, a dense core. Mature specific granules are 300 to 500 mμ in diameter and are composed of finely granular material that is not as dense as that of the azurophil granules. The myelocytes divide; and while specific-granule production continues, azurophil granule formation ceases, resulting in a greater proportion of specific granules in succeeding generations. When division stops and the nuclei become horseshoe-shaped, the cells are known as metamyelocytes. The mature cells are formed when the nuclei are segmented into lobes.

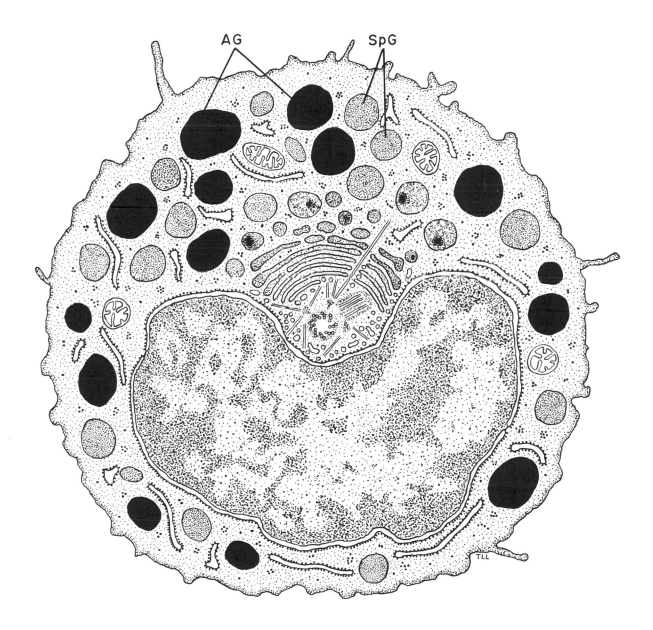

4 — POLYMORPHONUCLEAR LEUKOCYTE

The polymorphonuclear leukocytes comprise about 50 to 60 per cent of the leukocytes in the normal peripheral blood. The cell is 7 to 9 μ in diameter (10 to 12 μ in smears), with a few villi extending from its surface. The nucleus consists of three to five lobes joined by narrow chromatin threads. The chromatin material is condensed against the nuclear envelope, which is dilated to produce a perinuclear "halo." Most of the organelles are diminished in number. The cytoplasm contains relatively few mitochondria, cisternae of endoplasmic reticulum, ribosomes, and microtubules. Glycogen granules are numerous (Gly). The Golgi apparatus is not prominent, consisting of a few lamellae and vesicles adjacent to the nucleus. The centrioles may be absent. Many granules are distributed in the cytoplasm. About 80 to 90 per cent of the granules are specific (SpG), and 10 to 20 per cent are azurophil (AG). The specific granules are 300 to 500 mμ in diameter and are composed of material of medium density. Most are round, although a few rod or dumbbell shapes occur. The azurophil granules are dense and 600 to 800 mμ in diameter.

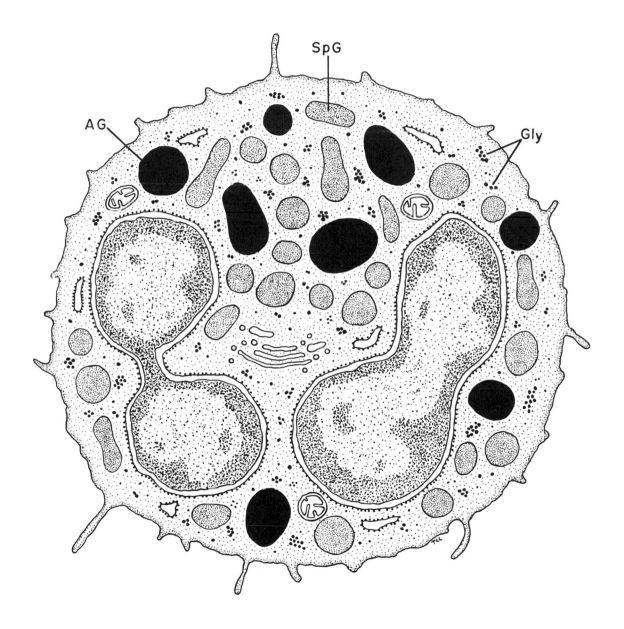

5 — POLYMORPHONUCLEAR LEUKOCYTE — PHAGOCYTOSIS

Polymorphonuclear leukocytes are active in the early stages of inflammation. When tissues are injured, chemical substances are released that influence the direction of movement of the leukocytes (chemotaxis). The cells migrate through the capillary walls into the tissues where they phagocytose bacteria and other particulate materials. By removing foreign substances and antigens, they play an important role in the defense mechanisms of the body. The illustrated cell is engulfing staphylococci. Cytoplasmic arms extend around a bacterium and fuse to incorporate it within a phagocytic vacuole (PhV). The granules of the polymorphonuclear leukocyte are lysosomes containing hydrolytic enzymes such as acid phosphatase, alkaline phosphatase, nucleotidase, ribonuclease, deoxyribonuclease, β-glucuronidase, cathepsin, and the antibacterial agents lysozyme and phagocytin. The membrane enclosing the granule fuses with the membrane of the vacuole containing the bacterium. In this manner, the enzymes comprising the granule are released into the vacuole and effect digestion of the bacterium. Because the enzymes are confined to the vacuole, the cell presumably is not sacrificed during this process. However, when the leukocytes are exposed to streptolysin, the granules are ruptured directly into the cytoplasm and cause lysis and death of the cell. Actively migrating cells protrude pseudopodia (Pd). These cytoplasmic extensions of the cell contain a few glycogen granules but are largely devoid of organelles.

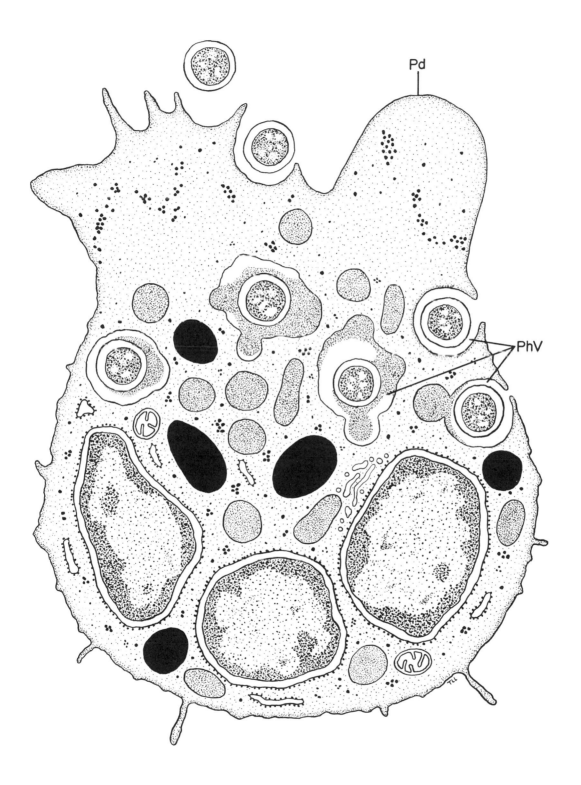

6—EOSINOPHILIC MYELOCYTE

The eosinophilic myelocyte, which arises from the progranulocyte (Plate 2), is a large rounded cell containing azurophil and specific granules in different stages of formation. The round nucleus contains small clumps of heterochromatin adjacent to the envelope and scattered in the nucleoplasm. A large Golgi zone composed of lamellae, vesicles, and vacuoles is found near the nucleus. Flocculent material is present in some of the lamellae and vacuoles. The vacuoles coalesce to form immature specific granules that undergo progressive condensation. Crystals (Cry) appear in some of the specific granules (SpG). At this stage, not all of the granules are ellipsoid, as in mature cells, but some granules are rounded. Non-specific or azurophil granules (AG) are relatively less abundant and are usually round and composed of dense material. Most of these granules are mature, but a few are incompletely condensed and have a less dense rim, which may contain vesicles, surrounding a dense core. In contrast to the other types of myelocytes, the eosinophilic myelocyte contains elongated cisternae of rough-surfaced endoplasmic reticulum. The cisternae are slightly dilated and contain a sparse amount of particulate material. Free ribosomes and large mitochondria are also found in the cytoplasm.

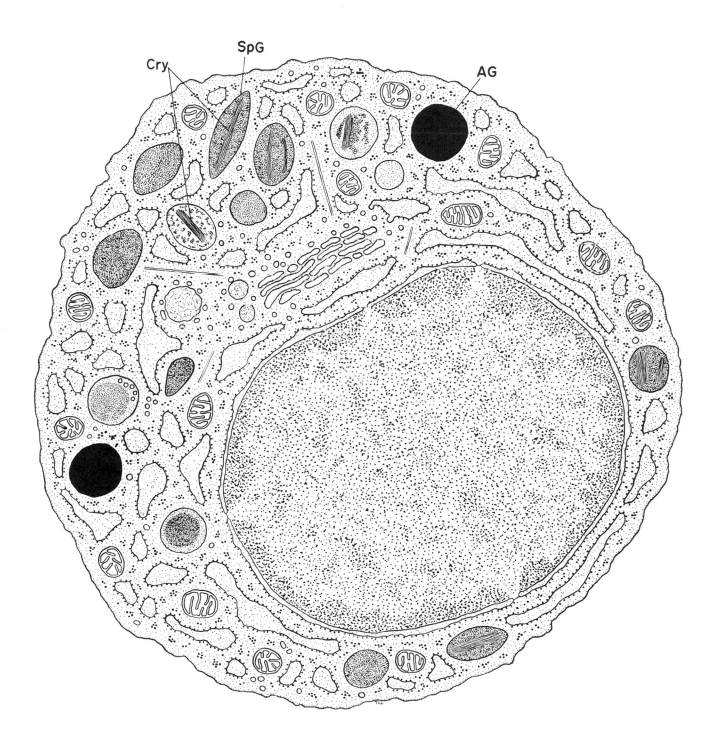

7 — EOSINOPHILIC LEUKOCYTE

Eosinophilic leukocytes form about 2 to 5 per cent of the total number of leukocytes in the normal peripheral blood. They are also found in the connective tissues. Eosinophils are round with a diameter of about 12 μ and may have a few surface villi or pseudopodia. The nucleus usually consists of two lobes joined by a chromatin strand. The chromatin material is densely packed, especially against the inner surface of the nuclear envelope. The cytoplasm contains a few elongated cisternae of rough-surfaced endoplasmic reticulum, free ribosomes, and small mitochondria. A Golgi apparatus, not very extensive, is adjacent to the nucleus and is composed of membranous lamellae and small vesicles.

Eosinophils are characterized by their numerous membrane-bounded granules. The granules are spherical, oval, or ellipsoid. They contain a matrix of medium density and often an equatorial band of extremely dense crystalloid material (Cry). The granules contain hydrolytic enzymes, but these granules differ from those of the polymorphonuclear leukocyte in their high content of peroxidase and their lack of the antibacterial agents lysozyme and phagocytin. The crystalline core of the granule may contain peroxidase. When materials are phagocytosed by the cell, the content of the granules is released and the ingested material is destroyed.

Eosinophilia is typical of allergic conditions such as parasitic infections and hypersensitivity reactions. Eosinophils appear to be attracted from the blood stream into the tissues by the presence of antigen-antibody complexes. During an antigen-antibody reaction, pharmacologically active compounds, including histamine, 5-hydroxytryptamine, and bradykinin, are released from other sources. One function of eosinophils is to limit the effects of these substances. In addition, they have been shown to phagocytose antigen-antibody complexes.

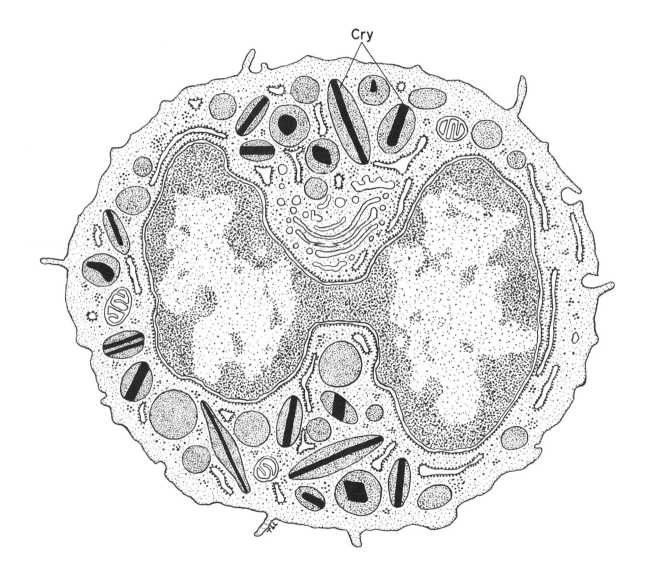

Cry

8—BASOPHILIC MYELOCYTE

The basophilic myelocyte is the next stage in differentiation of basophils from progranulocytes (Plate 2). This cell is smaller and less frequent than the other myelocytes and has a rounded nucleus that usually lacks a nucleolus. Some of the chromatin occurs in small clumps but, for the most part, is dispersed. Cisternae of rough-surfaced endoplasmic reticulum are short and few in number, but free ribosomes are abundant. Mitochondria are scattered through the cytoplasm. A pair of centrioles and a prominent Golgi apparatus constitute the centrosphere zone adjacent to the nucleus. In some of the vacuoles associated with the Golgi, small and extremely dense droplets are found. Large vacuoles contain particulate material with dense foci or tightly packed strands of dense material (cf. mast cell, Plate 27). These structures represent stages in specific-granule formation. Most of the granules in the basophilic myelocyte are large, spherical, and composed of opaque material.

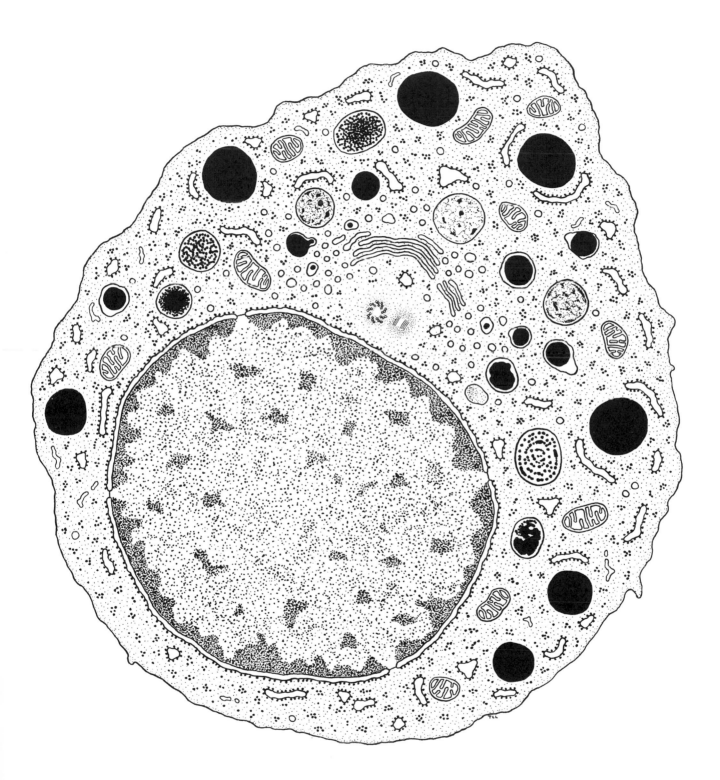

9 — BASOPHILIC LEUKOCYTE

Basophilic leukocytes comprise about 1 per cent of the number of leukocytes in the peripheral blood. The cell is round and has a diameter of 10 μ. The cytoplasm is filled with large spherical or, sometimes, ovoid granules. The granules have an internal structure that seems to differ from species to species or with different fixation procedures: crystalloid (Cry) and concentric lamellar structures have been found, but other granules are uniformly dense. The metachromatic, basophilic granules contain large amounts of heparin and histamine. The nucleus consists of lobes joined by strands. Chromatin is condensed adjacent to the nuclear envelope, and nucleoli are absent. Cytoplasmic organelles are few in number. There are some small mitochondria, a few short profiles of endoplasmic reticulum, a small number of free ribosomes, and some glycogen granules.

The basophils have structural and functional similarities to the mast cells (Plate 27), which also contain heparin and histamine. The granules of basophils are released in response to antigen-antibody reactions. Since heparin is an anticoagulant, the cells might function in inflammation to prevent clotting. Histamine causes dilation of small vessels and increased permeability of capillaries. This leads to increased diapedesis and, in turn, phagocytic activity by other leukocytes. Thus, these cells could serve to modify the inflammatory response, although their actual physiologic significance in this process has not been determined.

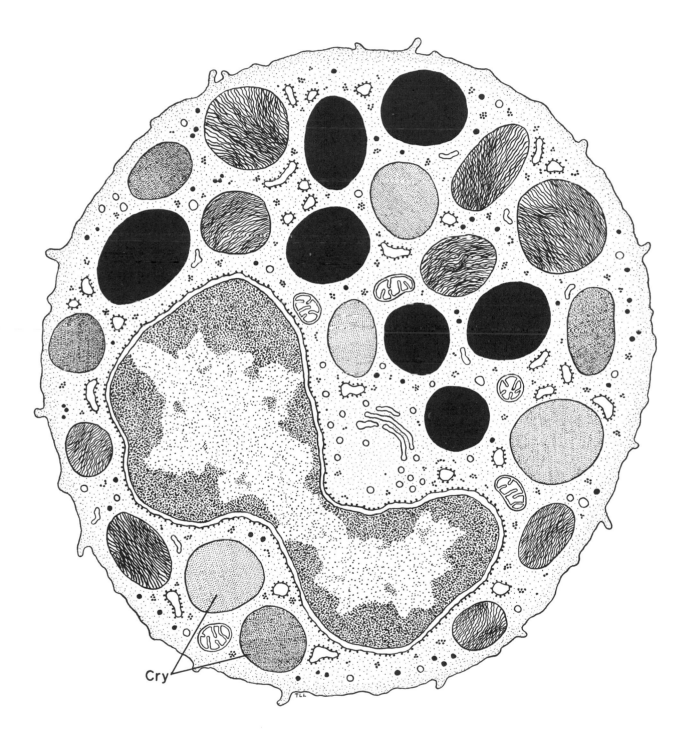

Cry

BLOOD, BONE MARROW, AND LYMPHATIC TISSUE

10—LYMPHOCYTE

Lymphocytes form 20 to 25 per cent of the leukocytes in the peripheral blood. They arise from and are found in the bone marrow, lymph nodes, thymus, spleen, tonsils, and lymphatic nodules in the lamina propria of the gastrointestinal tract. The circulating cells are spherical and 8 to 12 μ in diameter. Thin cytoplasmic extensions project from the surface. The nucleus is nearly spherical except for a small indentation on one side. The chromatin is highly condensed, and nucleoli are sometimes found in electron micrographs but not usually in smears. A relatively thin rim of cytoplasm surrounds the nucleus. A pair of centrioles and a small Golgi apparatus occur near the nuclear indentation. Mitochondria spherical to oval in shape are seen in the cytoplasm, especially around the Golgi region. Multivesicular bodies and isolated cisternae of rough-surfaced endoplasmic reticulum occur. Some lymphocytes contain a few dense granules (Gr) about 0.2 μ in diameter. These structures resemble lysosomes and may correspond to the azurophil granules observed in Romanowsky-stained dry smears.

A characteristic feature of lymphocytes is the presence of large numbers of ribosomes (R) free in the cytoplasm. Most of these occur singly, not as polyribosomes. Large lymphocytes have a greater amount of cytoplasm. Medium-sized lymphocytes are also found in the lymphatic tissues. These cells divide and give rise to small lymphocytes.

Lymphocytes play a major role in the immune response by serving as a source of antibody-producing cells. Upon making contact with an appropriate antigen in the lymph node, the small lymphocyte gives rise to a large dividing lymphocyte that differentiates into the plasma cell (Plate 19) responsible for antibody production. Lymphocytes may also be capable of transforming into monocytes and macrophages in the local inflammatory reaction.

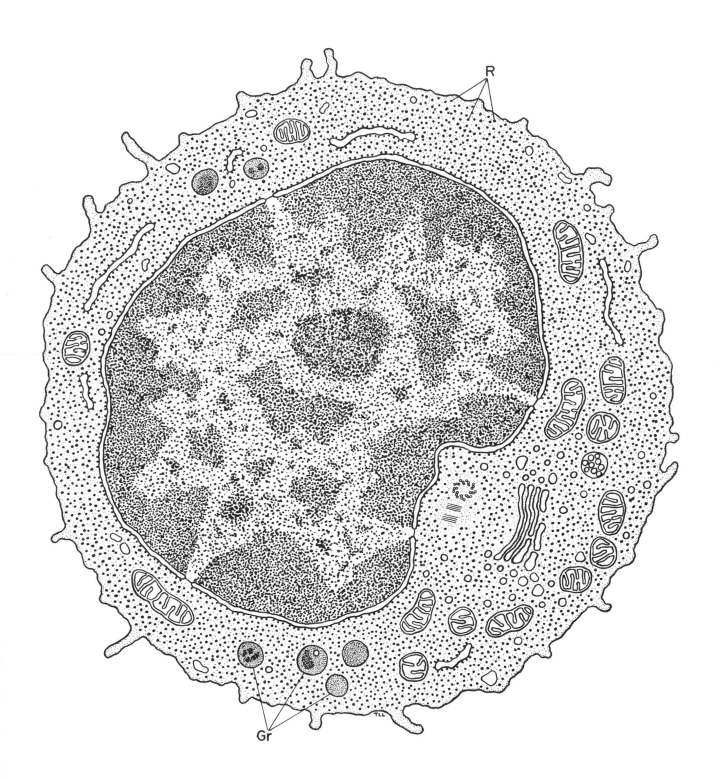

11—MONOCYTE

Monocytes are 9 to 12 μ in diameter (much larger in dry smears) and constitute 3 to 8 per cent of the leukocytes in the peripheral blood. The nucleus is horseshoe-shaped with clumps of condensed chromatin adjacent to the envelope and dispersed throughout the nucleoplasm. Nucleoli are not usually found in circulating monocytes. The cytoplasm is abundant. A pair of centrioles occurs near the nuclear indentation and is surrounded by stacks of Golgi cisternae. Many small vesicles are associated with this extensive Golgi apparatus. Vesicles (V) of various sizes are found throughout the cytoplasm. Monocytes contain a large number of small, dense, membrane-bounded granules (Gr) (lysosomes). Mitochondria are distributed randomly in the cytoplasm and have long, closely packed cristae and a dense matrix. Short cisternae of rough-surfaced endoplasmic reticulum and scattered free ribosomes are found. Sometimes an array of fine filaments is seen near the nucleus.

In the inflammatory reaction, monocytic cells appear in the tissues after the granulocytic cells. Monocytes leave the blood stream and enter the tissues, where they transform via an intermediate stage (polyblast) into actively phagocytic cells that remove bacteria, particulate material, and degenerating granulocytes. It is no longer possible to distinguish these cells, derived from monocytes, from the phagocytic cells (macrophages) that originate from fixed histiocytes. Thus, the blood monocyte represents a source of tissue macrophages (Plate 26).

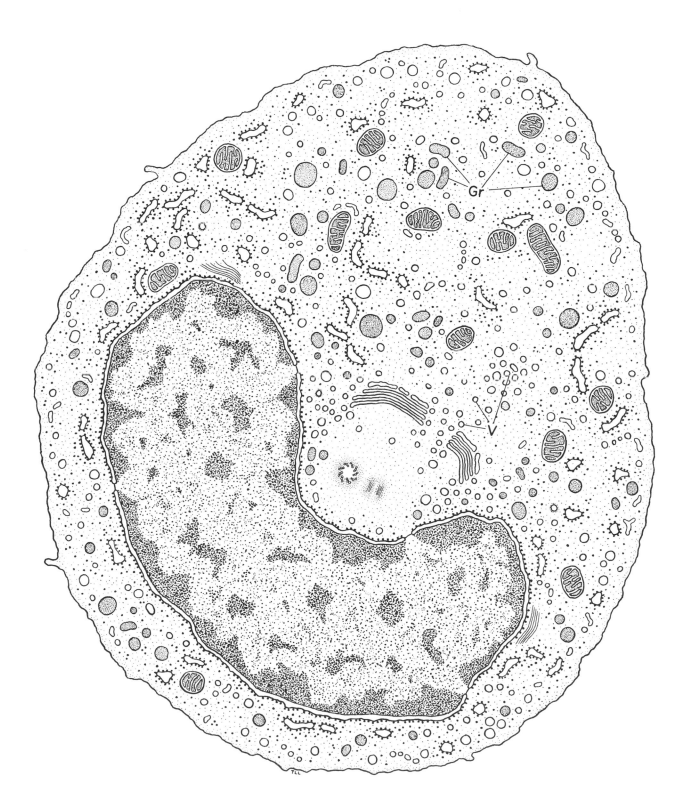

12 — BASOPHILIC ERYTHROBLAST

Erythropoiesis is the differentiation of erythrocytes from the stem cell or hemocytoblast (Plate 1). During this process, the cells undergo a series of changes related to the synthesis and accumulation of hemoglobin, and a simultaneous loss of ribonucleoprotein, the organelles, and the nucleus. Development of hemocytoblasts into erythroblasts is believed to be under the control of the hormone erythropoietin. The earliest stage in development is called the basophilic erythroblast (some investigators distinguish a proerythroblast stage). These cells divide and, like the stem cells, are active in DNA synthesis. RNA synthesis also occurs, and protein synthesis is increasing at this time.

The basophilic erythroblast is smaller (15 μ) than the hemocytoblast. The central nucleus is smaller, and clumping of the chromatin is more pronounced. Nucleoli have disappeared, or only fragments remain. Ribosomes (R) are abundant in the cytoplasm, often occurring in clusters of five or six. These polyribosomes, or polysomes, are engaged in hemoglobin synthesis. The abundance of ribonucleoprotein is responsible for the intense basophilia of the cytoplasm in Romanowsky preparations such as Wright's stain. Hemoglobin (Hb) appears as finely granular masses of material between the ribosomes. Mitochondria are present in the cytoplasm. The Golgi apparatus has disappeared except for a few vesicles near the nucleus. Only a few cisternae of endoplasmic reticulum remain. Microtubules (Mt) occur in the cytoplasm and are most numerous in a marginal band located a short distance below, and following the contour of, the cell surface. This marginal band of microtubules may maintain cell shape and form.

Some regions of the plasma membrane contain a coating of finely fibrillar material on the outer surface. This material occurs either in plaques along the surface or coating the pinocytotic vesicles (PV). Ferritin particles are attached to the fuzzy coating and are incorporated into the cell within pinocytotic vesicles. Ferritin does not adhere to other regions of the cell surface, indicating that the filamentous coating may represent a mechanism for selective uptake of ferritin. Ferritin serves as a source of iron that is utilized in the synthesis of hemoglobin. Other small vesicles with lucent contents occur near the surface.

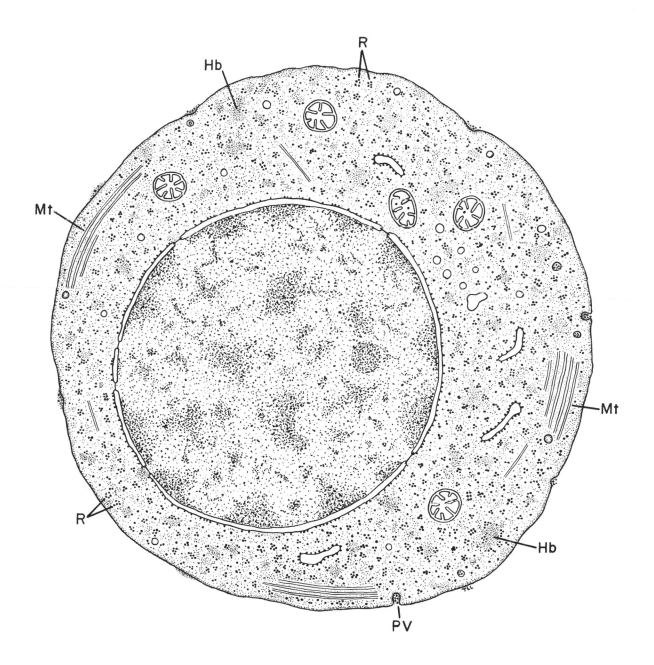

13 — POLYCHROMATOPHILIC ERYTHROBLAST

The polychromatophilic erythroblast is the next stage in the differentiation of erythrocytes from stem cells. There is no cell division beyond this stage. RNA synthesis decreases, but protein synthesis continues to increase. The cell is round and reduced in size (10 to 12 μ). The chromatin in the small round nucleus shows further condensation. Nucleoli are absent. Ribosomes occur in clusters in the cytoplasm but are not as numerous as they were previously. Fewer ribosomes occur on the surface of the nuclear envelope. The finely granular material thought to represent hemoglobin is much more abundant. At the light-microscopic level, the appearance of pink-staining hemoglobin in the basophilic cytoplasm produces a gray-green color; hence, the term polychromatophilic. Similar material occurs in the nucleus and is continuous with the cytoplasm through the nuclear pores. The ribosomes often are separated from the masses of hemoglobin by a thin region of low density. Mitochondria occur in the cytoplasm, but the Golgi apparatus has disappeared. Only a few small rough-surfaced cisternae of endoplasmic reticulum persist. The marginal band of microtubules (Mt) is illustrated in transverse section. Up to 18 microtubules comprise the bundle. Pinocytotic vesicles continue to form on the cell surface. Some of these vesicles and the larger vacuoles contain ferritin (Fe).

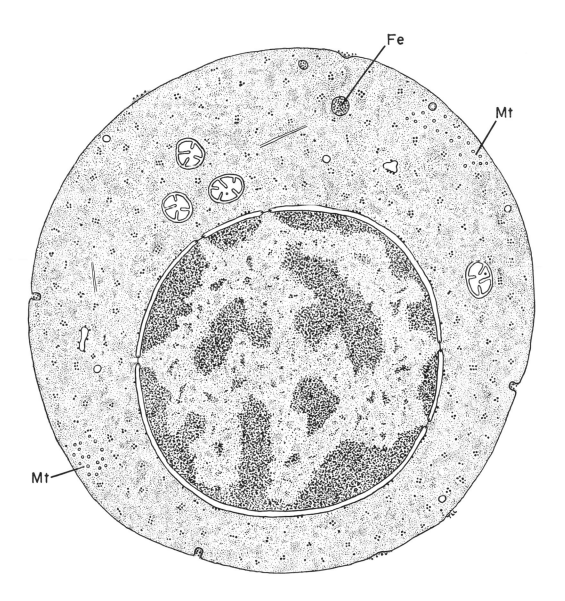

14—ORTHOCHROMATIC ERYTHROBLAST

The orthochromatic erythroblast, or normoblast, is about 8 to 9 μ in diameter, smaller than in previous stages. The nucleus is also smaller and may be eccentrically situated in the cell prior to its extrusion. The chromatin is very densely packed. Only a few nuclear pores remain at this stage. The cytoplasmic organelles are further reduced in number. A small number of mitochondria persist, but rough-surfaced cisternae and microtubules are rare. A few vesicles occur in the cytoplasm, and some contain ferritin particles. Pinocytotic invaginations continue to form on the surface but they are more infrequent than in previous stages. The cytoplasmic ribosomes are not as concentrated and show less tendency to occur in clusters. Hemoglobin, on the other hand, continues to accumulate, occupying greater regions of the cytoplasm. Clear areas may occur between the ribosomes and the masses of hemoglobin. The preponderance of hemoglobin in relation to RNA at this stage results in an orange color (orthochromatic) in stained preparations. Protein synthesis continues after RNA synthesis has ceased.

Denucleation begins with migration of the nucleus to one side of the cell and an accumulation of vesicles in the cytoplasm beneath the nucleus. The vesicles elongate and coalesce to produce larger spaces. These spaces fuse with the plasma membrane, freeing the nucleus and a thin rim of cytoplasm from the cell.

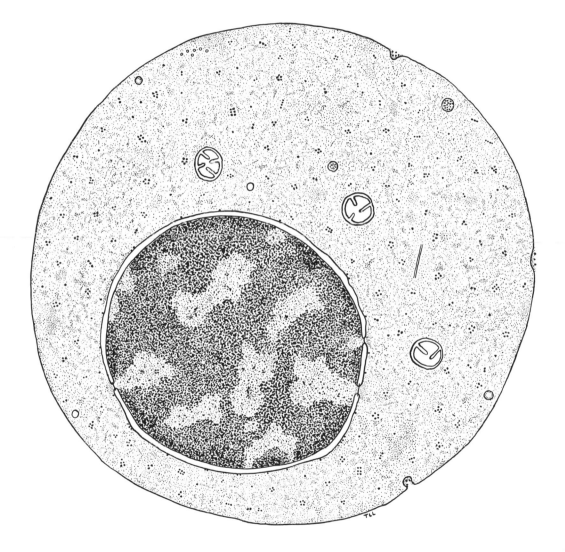

15—RETICULOCYTE

The reticulocyte has lost the nucleus and appears in the peripheral blood; it comprises 1 per cent of circulating erythrocytes. Reticulocytes show a further decrease in polyribosome content, although they are still active in protein synthesis. Sufficient ribonucleoprotein remains, so that the cell has a basophilic color when stained with Romanowsky stains. Supravital staining with brilliant cresyl blue precipitates the ribonucleoprotein into a web, or reticulum, thereby giving the cell its name. In immature reticulocytes, a few ribosomes are present and occur in clusters of two to six. As the cell matures, the total number of ribosomes is reduced, and the polyribosomes disaggregate into smaller clusters or single ribosomes. As ribosomes are lost, the cell loses its capacity for protein synthesis. The cytoplasm is filled with hemoglobin. A few mitochondria and vesicles persist, but other organelles are absent.

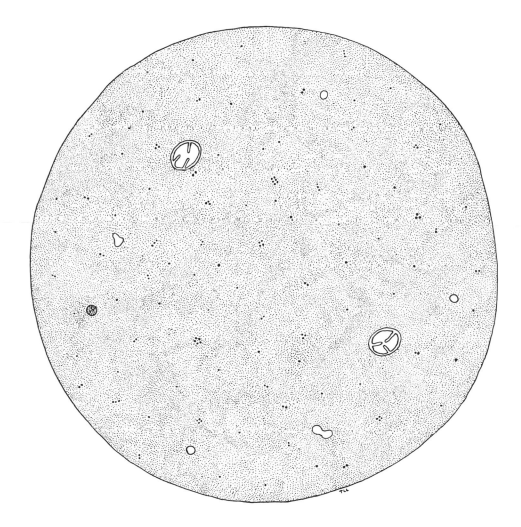

16—ERYTHROCYTE

The erythrocyte represents the final stage of erythrocytic differentiation. The mature cell is a biconcave disk about 7 to 8 μ in diameter. A membrane encloses the dense material comprising the cell. This material consists primarily of hemoglobin, the respiratory pigment; and its density is due to its iron content. In the mature cell, all traces of organelles have disappeared. Sometimes a few ribosomes persist in young erythrocytes, but the endoplasmic reticulum, mitochondria, microtubules, and vesicles are absent. These cells have lost their synthetic machinery and are incapable of growth and reproduction. Erythrocytes survive for about 120 days in the blood stream, after which they are phagocytosed by macrophages and reticular cells of the reticuloendothelial system in the spleen, liver, and bone marrow.

Hemoglobin forms 95 per cent of the dry weight of the erythrocyte and is concerned solely with the transport of oxygen. The erythrocyte is not inert, however, and it performs several functions requiring the expenditure of energy: it preserves its biconcave shape, maintains an electrolyte equilibrium by transporting potassium across the membrane into the cell and by excluding sodium, and it maintains hemoglobin in its reduced state. Energy in the form of ATP is supplied by utilization of glucose mainly via the Embden-Meyerhof glycolytic pathway and, to a lesser extent, by the oxidative pentose phosphate pathway. These pathways also provide, respectively, DPNH and TPNH, essential cofactors for the reduction of methemoglobin to hemoglobin by methemoglobin reductase.

17—MEGAKARYOCYTE

Megakaryocytes are giant cells (up to 40 μ) found in the bone marrow. They are the source of circulating platelets. Megakaryocytes arise from hemocytoblasts; and during differentiation, the nuclei but not the cytoplasm divide several times. Following each division, the daughter nuclei fuse to form the complex multilobed nucleus (lower left of Plate). The chromatin is condensed adjacent to the nuclear envelope, and nucleoli are numerous. Pairs of centrioles (Ce) are seen between the folds of the nucleus. Small Golgi complexes (G) are found throughout the cytoplasm. Associated with the Golgi and widely distributed in the cytoplasm are small, membrane-bounded dense granules (Gr). The endoplasmic reticulum is sparse, but free ribosomes are common. Small mitochondria are numerous.

Toward the periphery of the cell are extensive systems of paired membranes bounding a narrow cleft. These channels originate by the elongation and coalescence of previously discontinuous vesicles. The paired membranes, or platelet demarcation membranes or channels (PDC), become continuous with the cell surface, thereby separating fragments of cytoplasm from the cell proper. In this manner the thrombocytes, or platelets, are shed into the circulation.

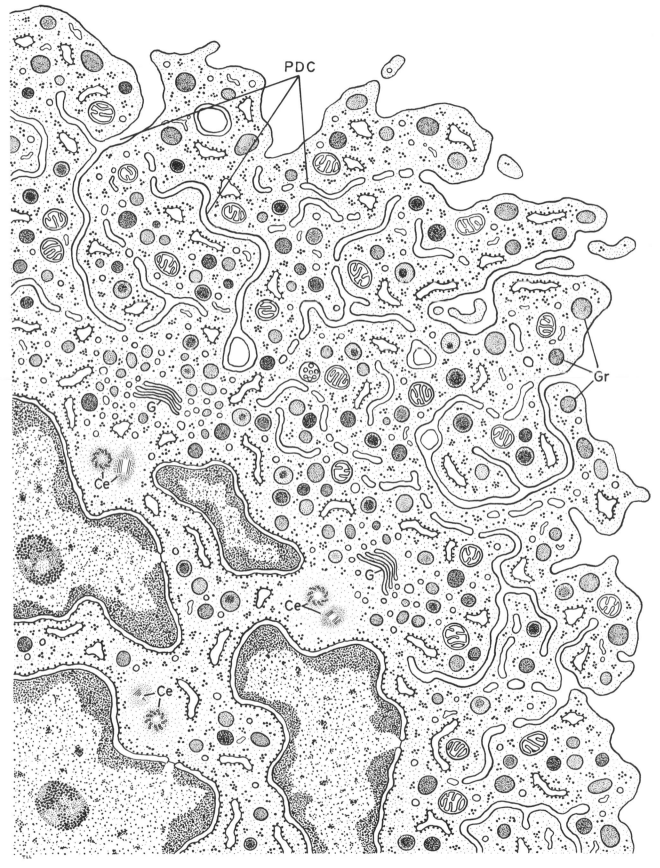

18—PLATELET

Fragments of megakaryocyte cytoplasm are shed into the circulation as platelets. Circulating platelets are small, biconcave disks, about 3 μ in diameter, that lack nuclei. The cytoplasmic content of platelets is similar to that of the megakaryocytes from which they originate. Dense membrane-bounded granules (Gr), small mitochondria, and short cisternae of endoplasmic reticulum are found. Other vacuoles (Vac) occur, some with lucent contents or with dense material surrounded by a clear halo. Small vesicles and glycogen granules are seen, but there are very few ribosomes. A marginal bundle of microtubules (Mt) is found beneath and parallel to the plasma membrane. These microtubules may be responsible for maintaining the discoid shape of the platelet.

Platelets have several roles in clot formation. Their surfaces are adhesive due to a layer of fibrinogen, and they agglutinate with one another and with the endothelium at the site of injury. In this manner they plug the defect in the vessel wall. Thromboplastin, which activates prothrombin, is also associated with the plasma membrane. The granules contain hydrolytic enzymes (acid phosphatase, β-glucuronidase) and may be lysosomes. Serotonin and histamine are also liberated by platelets.

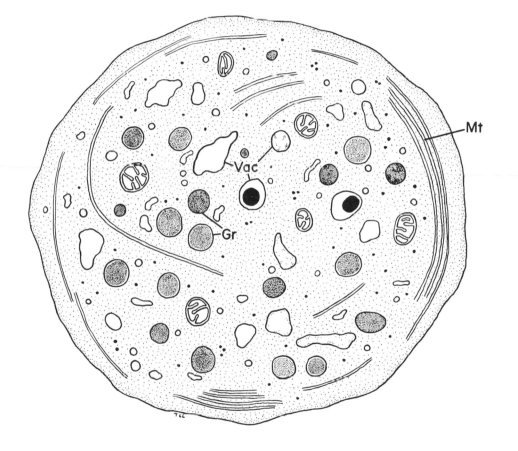

19 — PLASMA CELL

Plasma cells are found in lymphoid tissue and in small numbers in loose connective tissue. The cells are ovoid in shape, with a round, slightly eccentric nucleus. The chromatin material is condensed into large masses adjacent to the envelope and has a radial or cartwheel appearance on examination with the light microscope. Another large mass of condensed chromatin may occur in the center of the nucleus.

A large centrosome is found adjacent to the nucleus. A pair of centrioles occurs in the middle of this zone, and elements of the Golgi apparatus are peripheral. The Golgi complex is large and consists of flattened membranes, lamellae, and many small vesicles. On the inner face of the Golgi are some small granules composed of moderately dense material. The cytoplasm is filled with an extensive system of rough-surfaced endoplasmic reticulum. The cisternae are elongated, somewhat dilated, and contain amorphous material of medium density. Ribosomes, mitochondria, and occasionally a multivesicular body or lysosome are found between the cisternae.

Plasma cells are considered to be the major site of antibody formation. The extensive rough-surfaced endoplasmic reticulum is a characteristic feature of protein-secreting cells. Recently, an immunohistochemical technique has localized antibody material to the cisternae of the endoplasmic reticulum. Radioautography has shown that the polypeptide portion of the immunoglobulin molecule is synthesized in the endoplasmic reticulum and transported to the Golgi complex. Galactose is incorporated into the carbohydrate portion of the molecule in the Golgi complex, while incorporation of glucosamine takes place in both the endoplasmic reticulum and the Golgi. Unlike many cells synthesizing proteins for export (pancreatic exocrine cell, Plate 84), the plasma cell does not store its product in an extensive system of granules (cf., fibroblast, Plate 25). Most likely, vesicles derived from the Golgi apparatus and containing immunoglobulin migrate to the cell surface and fuse with it in a process of reverse pinocytosis.

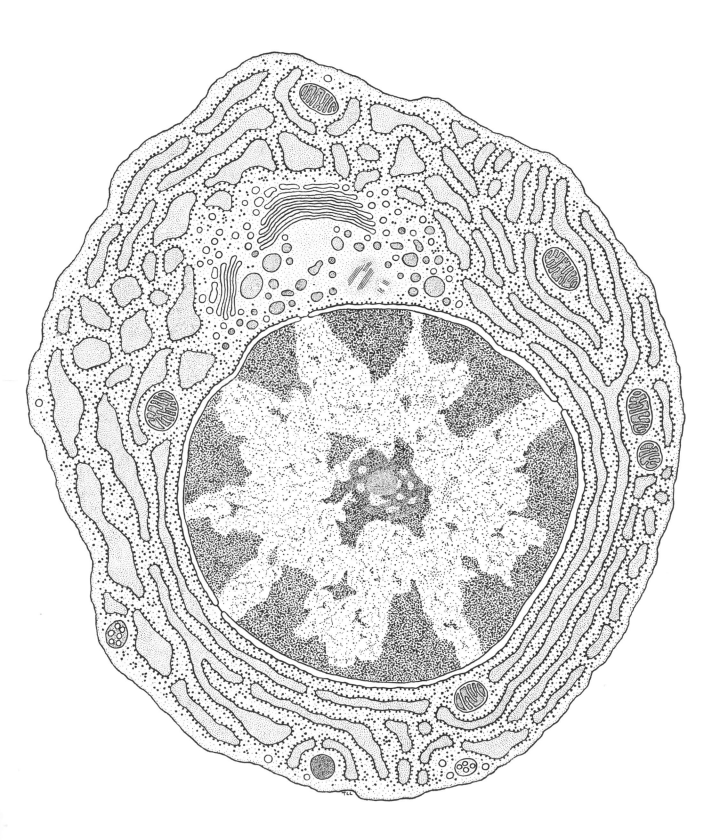

20—THYMUS: THYMOCYTE

Thymocytes are the most numerous cellular element in the thymus, especially in the cortex, where they are closely packed together. There seems to be little structural difference between the thymocytes and the lymphocytes (Plate 10) of other lymphoid organs or in the circulation. The cells have a large round nucleus and a thin rim of surrounding cytoplasm. The chromatin is condensed into large clumps, especially adjacent to the inside of the nuclear envelope. A nucleolus is seen in some cells. The nucleus bears an indentation at one pole, and opposite it is found a Golgi apparatus and a pair of centrioles. Mitochondria are distributed around the cell center and through the cytoplasm. A small number of lysosomes usually occur, also. There are only a few cisternae of endoplasmic reticulum, but free ribosomes are extremely abundant. The free ribosomes occupy most of the available cytoplasm and tend to occur in clumps or rosettes. The thymus plays a major role in the normal development of the immunological defense mechanisms. In young animals, thymocytes are produced in large numbers and seed the lymph nodes, spleen, and other lymphoid organs with lymphocytes. Having once acquired these cells, the lymphoid organs are immunologically competent and produce antibody to specific antigens.

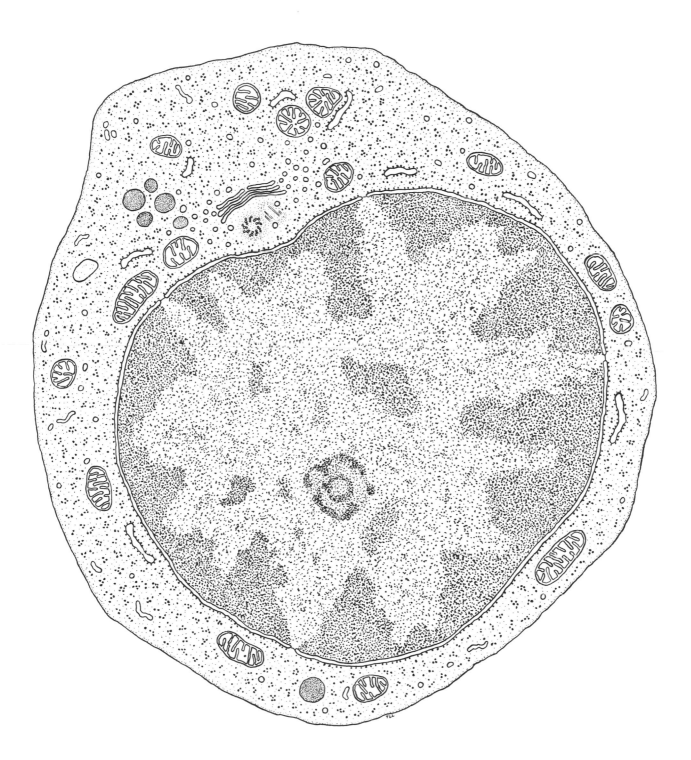

21 — THYMUS: EPITHELIAL RETICULAR CELL

The thymus is composed primarily of thymocytes (lymphocytes) and epithelial reticular cells. Epithelial reticular cells are of endodermal origin and can be differentiated from the phagocytic reticular cells of the reticuloendothelial system. Two basic types of epithelial cells can be distinguished, but these may be variations of a single cell type, since intermediate stages have been observed.

One type of epithelial cell, found in both the medulla and cortex, is highly irregular in shape with elongated cytoplasmic processes. These cells are wedged among lymphocytes and seem to form a supporting framework for the gland. Cytologically, these cells are relatively unspecialized and have a sparse distribution of cytoplasmic structures. The central nucleus contains dispersed chromatin material (in contrast to the lymphocytes), and a nucleolus. A small Golgi apparatus occurs near the nucleus. Mitochondria, a few short cisternae of rough-surfaced endoplasmic reticulum, and clusters of ribosomes are found. Some cells contain large dense granules that could represent secretory material, lysosomes, or keratohyalin granules. Bundles of tonofilaments (Tf) are conspicuous in the cytoplasm, and some terminate on the desmosomes (*maculae adherentes*, MA) that occur between adjacent epithelial cells.

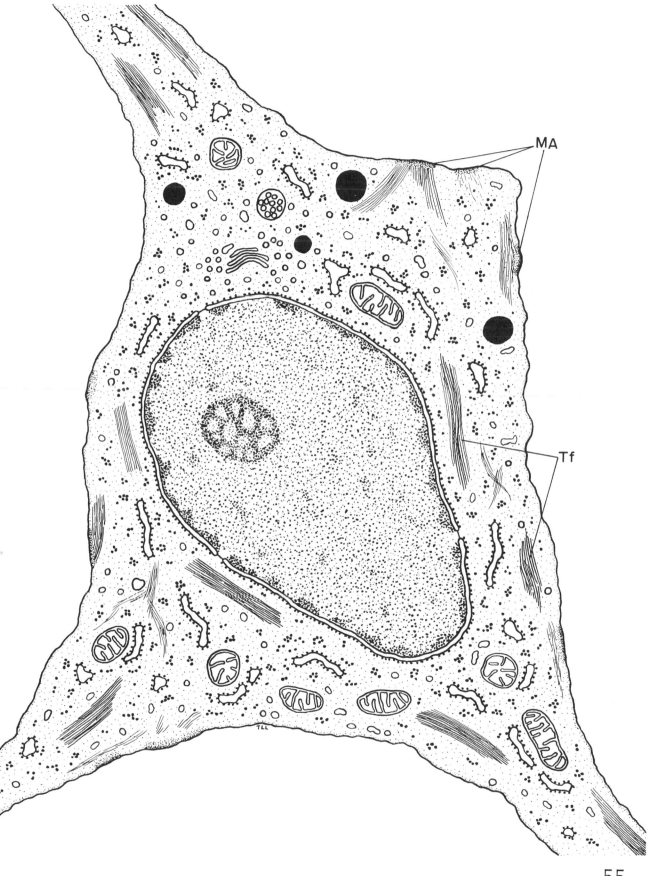

MA

Tf

22—THYMUS: EPITHELIAL RETICULAR CELL—MEDULLA

The second type of thymic epithelial cell is found in the medulla, where the ratio of epithelial cells to lymphocytes is greater than in the cortex. In the medulla, the cells are arranged in clumps or cords and vary in shape from irregular to round. The major difference from the other type of epithelial reticular cells is that these cells have a greater population of cytoplasmic structures which could be indicative of secretory activity. First is the occurrence of dense granules (Gr), the smallest appearing in the vicinity of the Golgi apparatus and larger ones distributed in the general cytoplasm. In addition, small vesicles with a less dense content are common near the Golgi. The vesicles appear to coalesce to form vacuoles (Vac). The vacuoles may be spherical and can contain varying amounts of particulate or amorphous material. Other vacuoles have contents of low density with microvilli extending into their cavity. Some of the vacuoles with villous protrusions may fuse and, in some cells, form a single, large, cystic cavity.

The nucleus of each cell is large, central in position, and contains a nucleolus. Short cisternae of rough-surfaced endoplasmic reticulum, ribosomes, and mitochondria are found in the cytoplasm. Filaments, organized in bundles of tonofibrils, are common and insert into desmosomes connecting adjacent cells.

The thymus contains a material that rapidly induces proliferation of lymphocytes in other lymphoid organs. The production of this material has been associated with the epithelial cells of the thymic medulla because of their apparent secretory activity. Which of the cytoplasmic structures (vesicles, granules, vacuoles) might contain the supposed hormonal factor, however, is not known.

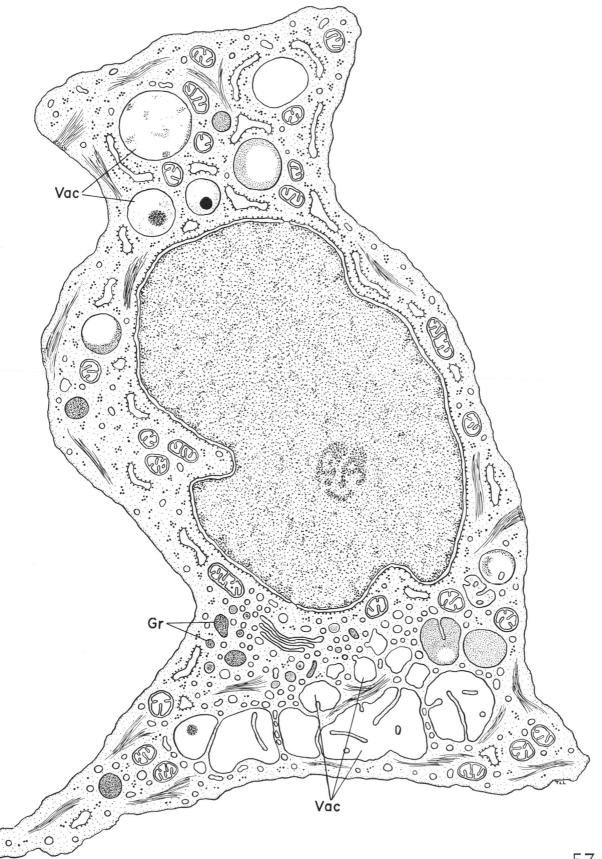

Vac

Vac

Gr

23—THYMUS: HASSALL'S CORPUSCLE EPITHELIAL CELL

Hassall's corpuscles are rounded structures composed of concentrically arranged cells located in the medulla of the thymus. The nature of these bodies has been disputed, but the electron microscope shows that they are composed of stages of keratinizing, stratified squamous epithelium. On the outside of the corpuscle are found relatively undifferentiated cells similar to the thymic epithelial reticular cells and possibly homologous with the basal cells of the epidermis. Above these cells and toward the center of the corpuscle are stages similar to cells of the stratum spinosum and the stratum germinativum and, finally, to the flattened keratinized cells of the stratum corneum.

An example of an intermediate stage comparable to a granular cell of the epidermis (Plate 50) is illustrated here. The most conspicuous elements in the cytoplasm are the electron-dense keratohyalin granules (KG) and the tonofilaments (Tf) aggregated into fibrils. The keratohyalin granules are usually contiguous with the tonofibrils. The tonofibrils also insert on the numerous desmosomes associated with the cell surface.

THYMUS: HASSALL'S CORPUSCLE EPITHELIAL CELL

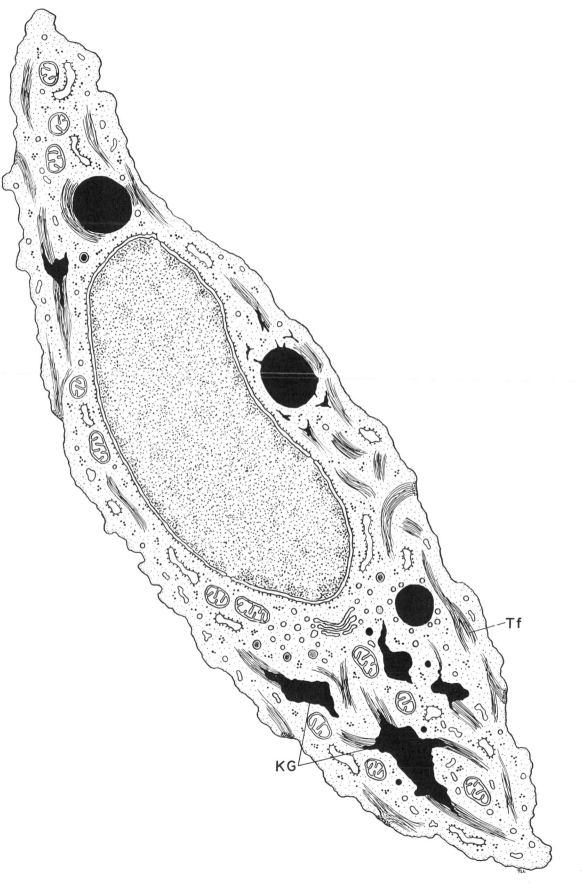

Tf

KG

BLOOD, BONE MARROW, AND LYMPHATIC TISSUE

24 — GLOBULE LEUKOCYTE

The globule leukocyte is found lying between epithelial cells in the mucous membranes of the digestive tract, respiratory tract, and urinary tract. It is characterized by its globular, acidophilic inclusions. The cells are roughly spherical with surface protrusions and filopodia. The nucleus is eccentrically located and contains a nucleolus and patches of condensed chromatin. The globules or granules are large, 0.5 to 1.5 μ in diameter, and bounded by a membrane. The content of the granules is variable. Most are composed of finely granular, dense material. Others are more mottled in appearance. Some show a coarsely granular cortex around a more homogeneous core. Other granules contain crystalline inclusions or fibrous bodies. Fibrous bodies have also been observed free in the hyaloplasm. A Golgi apparatus enclosing a pair of centrioles is situated near the nucleus. There are a few short cisternae of rough-surfaced endoplasmic reticulum, ribosomes, and mitochondria scattered in the cytoplasm.

Both the origin and function of globule leukocytes are unclear. It has been suggested that they arise by transformation from other cells such as lymphocytes, plasma cells, and mast cells. A correlation does appear to exist between the number of globule leukocytes and the presence of parasites and certain inflammatory conditions. These cells could therefore be involved in an immunological defense mechanism.

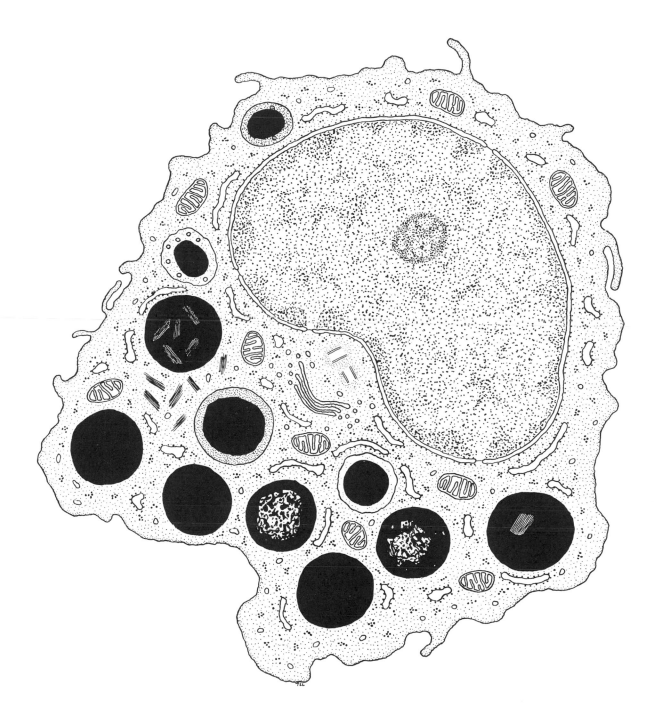

REFERENCES

Anderson, D. R., 1966. Ultrastructure of normal and leukemic leukocytes in human peripheral blood. J. Ultrastruct. Res., Suppl. 9, 1-42.

Bainton, D. F. and M. G. Farquhar, 1966. Origin of granules in polymorphonuclear leukocytes. Two types derived from opposite faces of the Golgi complex in developing granulocytes. J. Cell Biol., 28:277-302.

Bainton, D. F. and M. G. Farquhar, 1970. Segregation and packaging of granule enzymes in eosinophilic leukocytes. J. Cell Biol., 45:54-73.

Clark, S. L., 1963. The thymus in mice of strain 129/J, studied with the electron microscope. Amer. J. Anat., 112:1-34.

Fawcett, D. W., 1965. Surface specializations of absorbing cells. J. Histochem. Cytochem., 13:75-91.

Fedorko, M. E. and J. G. Hirsch, 1965. Crystalloid structure in granules of guinea pig basophils and human mast cells. J. Cell Biol., 26:973-976.

Grasso, J. A., 1966. Cytoplasmic microtubules in mammalian erythropoietic cells. Anat. Rec., 156:397-414.

Grasso, J. A., H. Swift, and G. A. Ackerman, 1962. Observations on the development of erythrocytes in mammalian fetal liver. J. Cell Biol., 14:235-254.

Hirsch, J. G. and M. E. Fedorko, 1968. Ultrastructure of human leukocytes after simultaneous fixation with glutaraldehyde and osmium tetroxide and "postfixation" in uranyl acetate. J. Cell Biol., 38:615-628.

Ito, T. and T. Hoshino, 1966. Fine structure of the epithelial reticular cells of the medulla of the thymus in the golden hamster. Z. Zellforsch., 69:311-318.

Kent, J. F., 1966. Distribution and fine structure of globule leucocytes in respiratory and digestive tracts of the laboratory rat. Anat Rec., 156:439-454.

Mandel, T., 1968. The development and structure of Hassall's corpuscles in the guinea pig. A light and electron microscopic study. Z. Zellforsch., 89:180-192.

Murray, R. G., A. Murray, and A. Pizzo, 1965. The fine structure of the thymocytes of young rats. Anat. Rec., 151:17-40.

Rifkind, R. A., D. Danon, and P. A. Marks, 1964. Alterations in polyribosomes during erythroid cell maturation. J. Cell Biol., 22:599-611.

Silver, M. D., 1965. Cytoplasmic microtubules in rabbit platelets. Z. Zellforsch., 68:474-480.

Takeuchi, A., H. R. Jervis, and H. Spring, 1969. The globule leucocyte in the intestinal mucosa of the cat: A histochemical, light and electron microscopic study. Anat. Rec., 164:79-100.

Wetzel, B. K., R. G. Horn, and S. S. Spicer, 1967. Fine structural studies on the development of heterophil, eosinophil, and basophil granulocytes in rabbits. Lab. Invest., 16:349-382.

Zagury, D., J. W. Uhr, J. D. Jamieson, and G. E. Palade, 1970. Immunoglobulin synthesis and secretion. II. Radioautographic studies of sites of addition of carbohydrate moieties and intracellular transport. J. Cell Biol., 46:52-63.

Zucker-Franklin, D. and J. G. Hirsch, 1964. Electron microscope studies on the degranulation of rabbit peritoneal leukocytes during phagocytosis. J. Exp. Med., 120:569-576.

CONNECTIVE TISSUE

CONNECTIVE TISSUE

25 — FIBROBLAST

Fibroblasts are the common fixed cells of connective tissue. They occur in loose areolar connective tissue and in dense connective tissue including the dermis, fascia, tendons, ligaments, and cornea. These cells are thought to be responsible for the formation of the connective tissue fibers and amorphous ground substance. Fibroblasts are spindle-shaped or stellate, and have an oval nucleus. Nucleoli are present, but condensed chromatin is sparse, occurring in small clumps adjacent to the nuclear envelope and dispersed in the nucleoplasm. A pair of centrioles and a small Golgi apparatus (G) are situated near the nucleus. Some small vacuoles (Vac) in the Golgi region contain flocculent material that may represent precursors of collagen or other extracellular substances produced by fibroblasts. Clusters of ribosomes and some short cisternae of rough-surfaced endoplasmic reticulum (ER) are found in the cytoplasm. In actively secreting cells, the Golgi apparatus and endoplasmic reticulum are more extensive. Mitochondria vary in shape from oval to elongated. Cytoplasmic tonofilaments (Tf) are abundant, and a few lysosomes may be found. Some small vesicles are associated with the cell surface.

Collagen synthesis and secretion follow a pathway basically similar to that of other secretory cells except that there is not extensive storage of the synthetic product in granules. Proline, a major constituent of collagen, enters the cell and is hydroxylated to hydroxyproline. Hydroxyproline and other amino acids are assembled into peptides on the ribosomes of the endoplasmic reticulum and are released into the cisternae. These peptides are then incorporated into three alpha chains that coil together in a helical fashion to form the tropocollagen molecule. Tropocollagen is about 2800 Å long and 14 Å wide and is the basic molecular unit of collagen. It is segregated in vacuoles in the Golgi complex. The vacuoles move to the cell surface and fuse with the plasma membrane, liberating their contents. Polymerization of tropocollagen molecules extra-cellularly produces the collagen fibrils (Co). The typical cross banding of collagen fibrils at intervals of 640 Å is produced by the overlap of about one quarter of the length of the laterally associated tropocollagen molecules. The mucopolysaccharides of the ground substance may be synthesized in the Golgi complex.

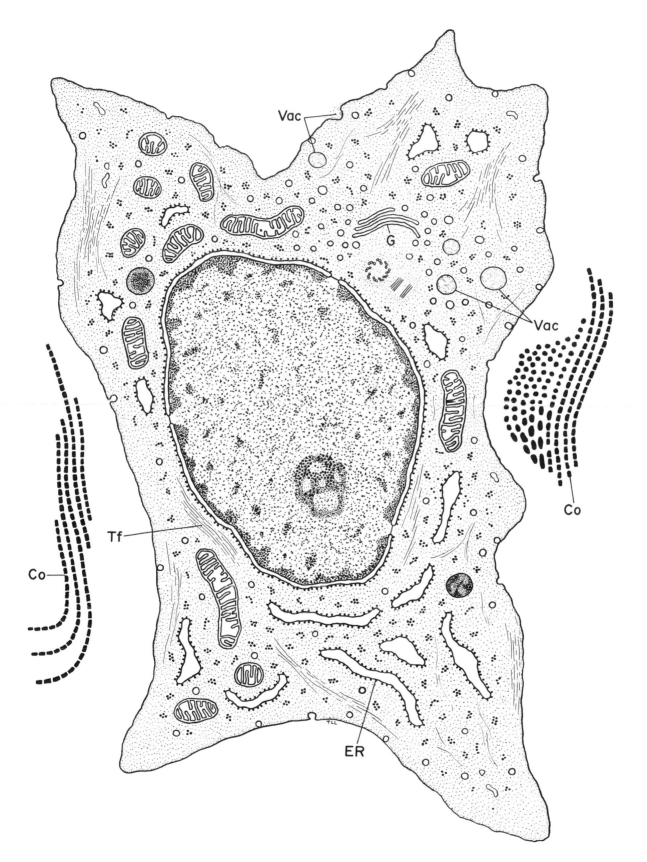

CONNECTIVE TISSUE

26—MACROPHAGE

Macrophages, or histiocytes, are cells of the connective tissue that are capable of becoming motile and phagocytic. They have a round central nucleus containing a nucleolus. A pair of centrioles occurs near the nucleus and is flanked by lamellae of the Golgi apparatus. Small vesicles and dense granules are abundant in the Golgi zone. Larger granules (lysosomes, Ly) composed of moderately to extremely dense material are found throughout the cytoplasm. Phagocytic vacuoles (PhV) are larger and have a heterogeneous content of particulate or granular material, myelin figures, or crystalline material. Filaments, sometimes in bundles, run through the cytoplasm. Lipid droplets (LD), short cisternae of rough-surfaced endoplasmic reticulum, ribosomes, and mitochondria complete the types of cytoplasmic structures. A narrow zone of surface ectoplasm, pseudopodia (Pd), and phagocytic arms are largely devoid of organelles. A few pinocytotic vesicles occur on the cell surface.

Migratory phagocytic macrophages are derived from fixed tissue histiocytes and blood monocytes (Plate 11). The lysosomes of macrophages contain nucleases, proteinases, and carbohydrases but, unlike granulocytes, are rich in lipases. Lipase is active against the lipoid capsule of the mycobacteria of tuberculosis and leprosy. During the process of phagocytosis and destruction of the tubercle bacillus, the macrophages are transformed into epithelioid cells. They may also undergo nuclear division in the absence of cytoplasmic division and produce giant, multinucleated cells.

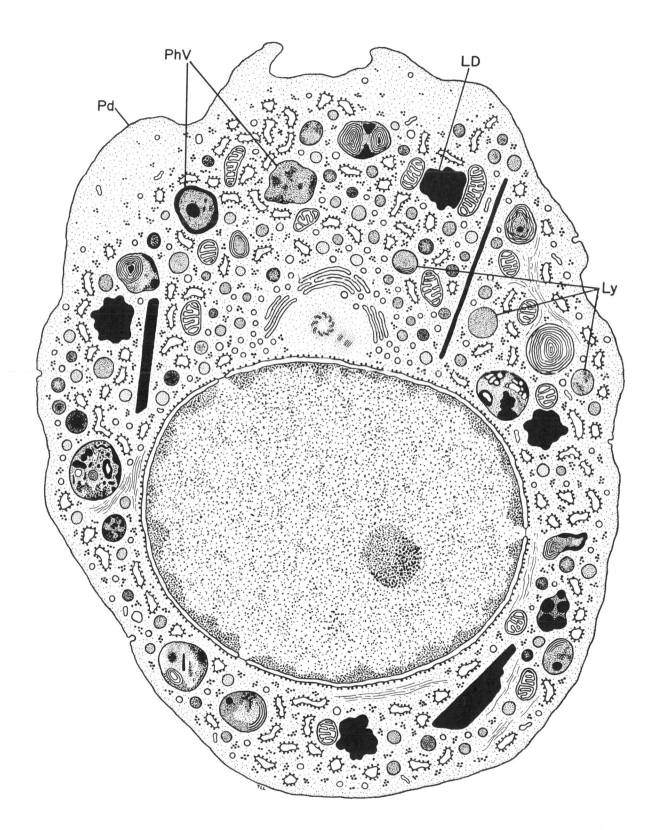

PhV

Pd

LD

Ly

CONNECTIVE TISSUE

27 — MAST CELL

Mast cells are round or ovoid cells found in connective tissue especially along the small blood vessels. The central nucleus conforms to the shape of the cell and is small in relation to the cytoplasmic space. Most of the cytoplasmic space is occupied by large membrane-bounded granules. The granules are extremely dense and may have a crystalline substructure (Cry). Immature granules near the Golgi apparatus have a less dense matrix within which is embedded irregular aggregates or strands of extremely dense material. Organelles are few in number and are compressed in the small amount of cytoplasm between the granules. Present are a few short cisternae of rough-surfaced endoplasmic reticulum, mitochondria, and ribosomal clusters. A Golgi apparatus (G) is found near the nucleus. Vacuoles associated with the Golgi apparatus contain dense droplets. Some short microvilli extend from the cell surface.

Mast cell granules contain heparin, histamine, and, in some species, 5-hydroxytryptamine or serotonin. Thus, these cells may modify local inflammatory reactions by influencing the blood-clotting mechanism, vasoconstriction, and capillary permeability. Similar functions have been suggested for the basophil (Plate 9), which also contains heparin and histamine. There appears to be a close functional connection between mast cells and basophils, although they probably represent two separate cell systems. Mast cells are also believed to produce hyaluronic acid and, because of the lipid-clearing activity of heparin, to play a role in regulation of lipid metabolism.

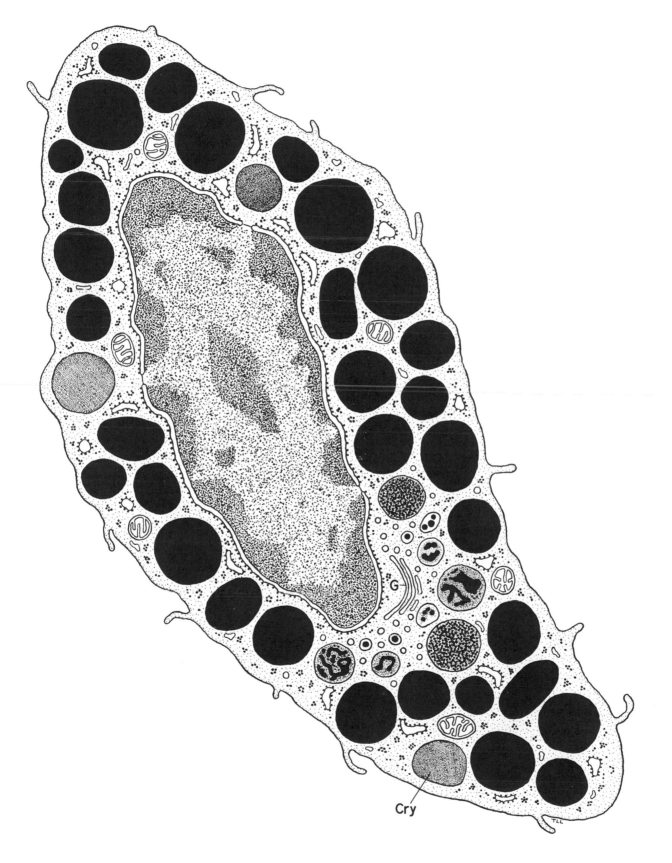

Cry

CONNECTIVE TISSUE

28—WHITE ADIPOSE CELL

White or ordinary adipose tissue occurs in the subcutaneous layer (panniculus adiposus) and in the omentum, mesenteries, and retroperitoneal regions. Isolated fat cells are also present in loose connective tissue. White adipose cells are large with a spherical or polyhedral shape. Mature cells have a large central lipid droplet surrounded by a thin peripheral rim of cytoplasm, giving the cell its signet-ring appearance. The lipid droplet appears homogeneously opaque after osmium tetroxide fixation; it is not bounded by a membrane. The nucleus occupies a peripheral position and is compressed to a crescent shape. Organelles in the rim of cytoplasm are mitochondria, which are spherical with a dense matrix and with cristae that extend across the organelle; an inconspicuous Golgi apparatus; a few short cisternae of rough-surfaced endoplasmic reticulum; and scattered profiles of smooth vesicles or tubules. Some of the latter structures may represent elements of smooth-surfaced endoplasmic reticulum. Some free ribosomes are present in the cytoplasm, and pinocytotic invaginations are common at the cell surface.

Fat cells are more than static storage depots: they are constantly undergoing a turnover in their lipid content. During lipid accumulation, fatty acids produced by hydrolysis of triglycerides in the chylomicrons originating from the intestine (see Plate 77) or derived from serum lipoproteins are transported into the cell, or triglycerides are synthesized from carbohydrate within the cell. Lipid is stored in the form of neutral triglycerides, and it represents the organism's main storage form of available energy. The lipid stores are mobilized through the action of lipase on triglycerides, yielding fatty acids that are released from the cell. Lipid assimilation and mobilization is under hormonal and nervous control.

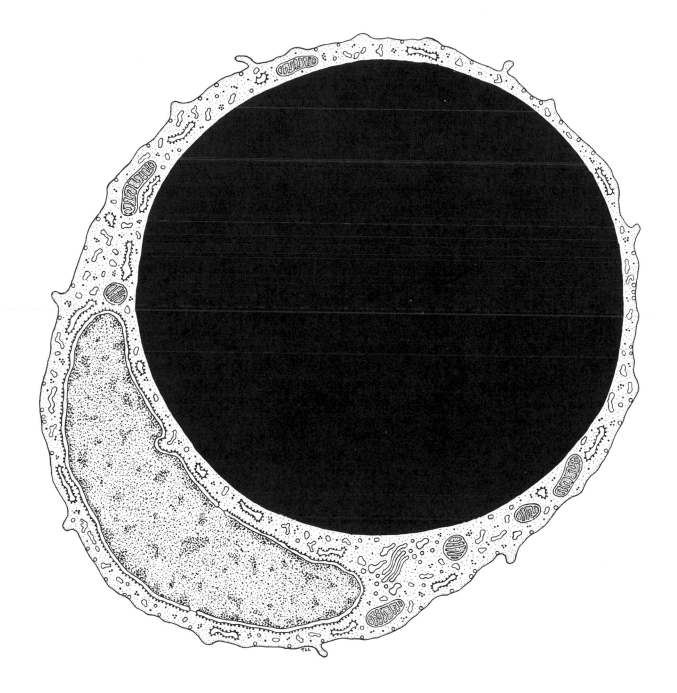

29—BROWN ADIPOSE CELL

Brown fat is not as widely distributed as white adipose tissue. It is found in restricted locations such as the interscapular region of rodents and hibernating animals and in the axillae of monkeys. The cells differ from those of white fat, being smaller and polygonal in shape and containing multiple small lipid droplets (LD). The lipid droplets are not enclosed by a membrane except for a few of the smallest ones. The nucleus is eccentrically located and often contains two nucleoli (Nl). The amount of cytoplasm in relation to lipid is greater in brown fat than in white fat. Much of the cytoplasm is occupied by numerous large mitochondria (M). Cristae are abundant and extend across the organelle. The mitochondrial matrix is moderately dense, but opaque granules are not prominent. The hyaloplasm is of low density and contains free ribosomes and glycogen granules. Smooth- and rough-surfaced cisternae of endoplasmic reticulum are sparse. A small Golgi apparatus is present. Some lysosomes occur in the cytoplasm, and pinocytotic vesicles are associated with the surface.

Brown fat is thought to serve as a source of heat in animals emerging from hibernation. Mobilization of fatty acids is controlled by the nervous system. Norepinephrine is released at nerve endings, and this leads to the activation of lipase in the fat cells. Lipase effects the breakdown of triglyceride to fatty acids and glycerol. Oxidation or re-esterification of fatty acid is accompanied by oxygen consumption and generation of heat that warms the blood passing through the brown adipose tissue.

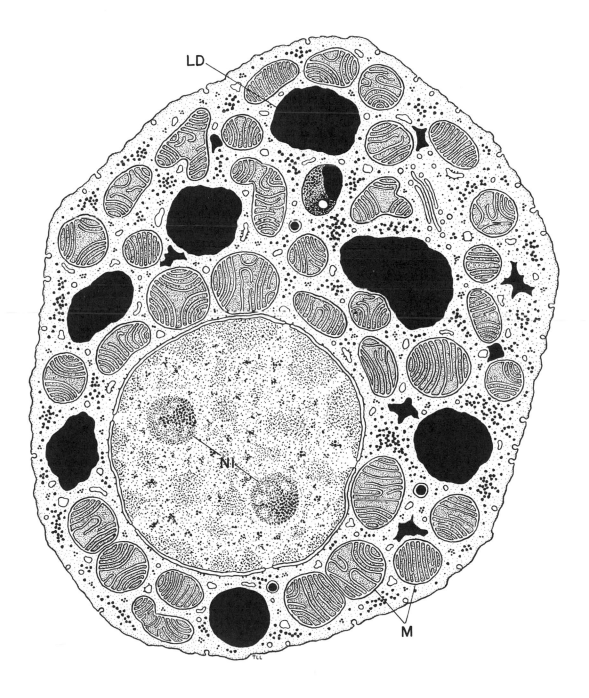

CONNECTIVE TISSUE

30—CHONDROCYTE

Chondrocytes, or cartilage cells, lie within lacunae in the matrix of cartilage. The cells are round to oval in profile, with slender processes projecting from the surface. The large central nucleus may contain a nucleolus. Chondrocytes secrete the collagen and the chondromucoprotein of the cartilage matrix and have the usual complement of organelles associated with protein synthesis. These structures include elongated cisternae of rough-surfaced endoplasmic reticulum and a Golgi apparatus. The latter is found near the nucleus and consists of lamellae, vesicles, and vacuoles (Vac). The vacuoles often assume the shape of dilated saccules and contain flocculent material. Some of the vacuoles are situated near the cell surface. Mitochondria and free ribosomes also occur in the cytoplasm. Glycogen granules (Gly) and lipid droplets (LD) accumulate in cells that are not actively secreting.

The cartilage matrix consists of an amorphous ground substance within which are embedded collagen fibrils. These fibrils actually constitute over 40 per cent of the dry weight of cartilage. Chondrocytes synthesize and release tropocollagen in the same manner as the fibroblasts (Plate 25). They are also responsible for the elaboration of the ground substance whose principal constituent is chondromucoprotein, a polymer of chondroitin sulfate and a mucoprotein. The polysaccharide components of the mucoprotein are probably synthesized in the Golgi complex; sulfation of mucopolysaccharides occurs here as well. The protein is synthesized in the endoplasmic reticulum and is then transferred to the Golgi region. Here it combines with the polysaccharide and is secreted as chondromucoprotein.

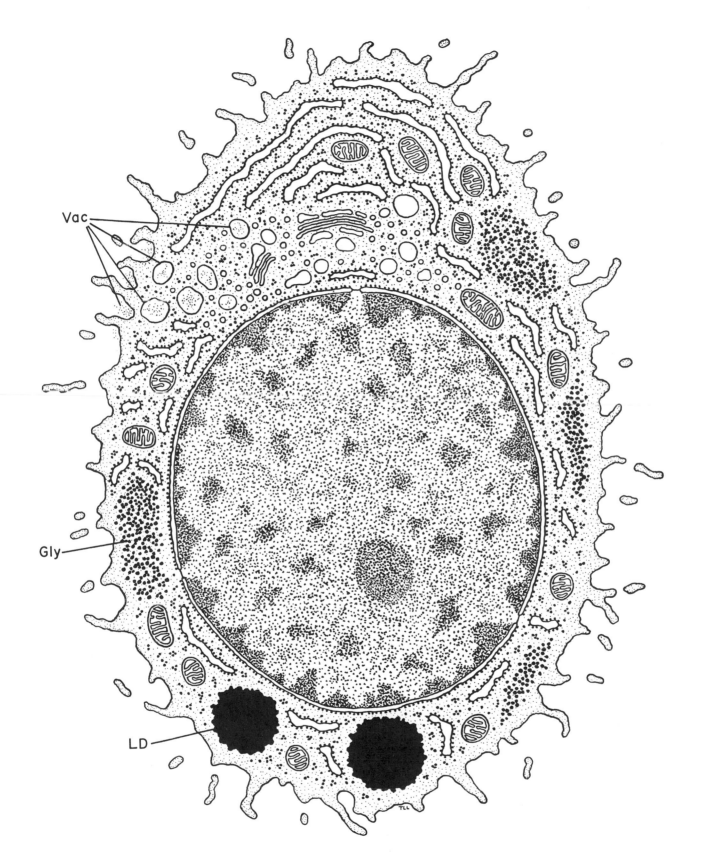

Vac

Gly

LD

CONNECTIVE TISSUE

31 — OSTEOCYTE

Osteocytes lie within lacunae of fully formed bone. The surrounding bone matrix consists of a mucopolysaccharide ground substance, collagen fibers, and hydroxyapatite crystals. When the bone matrix is completely mineralized, as shown here, it is opaque. A space containing some collagen fibrils (Co) is found between the cell membrane and the mineralized matrix. The osteocyte is flattened but has slender cytoplasmic processes extending into canaliculi (Ca) within the bone matrix. The nucleus conforms to the shape of the cell. A Golgi apparatus, consisting of membranous lamellae, small vesicles, and larger vacuoles, is found adjacent to the nucleus. Elongated, parallel, membranous cisternae with ribosomes on their outer surfaces comprise the endoplasmic reticulum. Free ribosomes are found in the hyaloplasm. Other cytoplasmic organelles are spherical mitochondria and lysosomes. Cells that actively secrete bone matrix have a more extensive Golgi apparatus and endoplasmic reticulum (osteoblasts), but these organelles are less prominent in quiescent cells that have been encased by bone material (osteocytes). Osteocytes are not inactive, however, and play a role in the maintenance of the bone matrix.

The bone matrix consists of organic and inorganic components. The organic matrix is composed of collagen fibers and an amorphous ground substance, a mucopolysaccharide. These components of the matrix are synthesized by osteocytes in a manner presumably the same as that occurring in fibroblasts (Plate 25) and chondrocytes (Plate 30). Once the organic matrix is formed, it becomes mineralized, imparting great hardness to the bone. The inorganic material is hydroxyapatite $[Ca_{10}(PO_4)_6(OH)_2]$. The hydroxyapatite crystals are slender needles that form on the collagen fibers. It is thought that initial crystal deposition is governed by the macromolecular structure of collagen, possibly an enzyme located at a particular site on the collagen fibril.

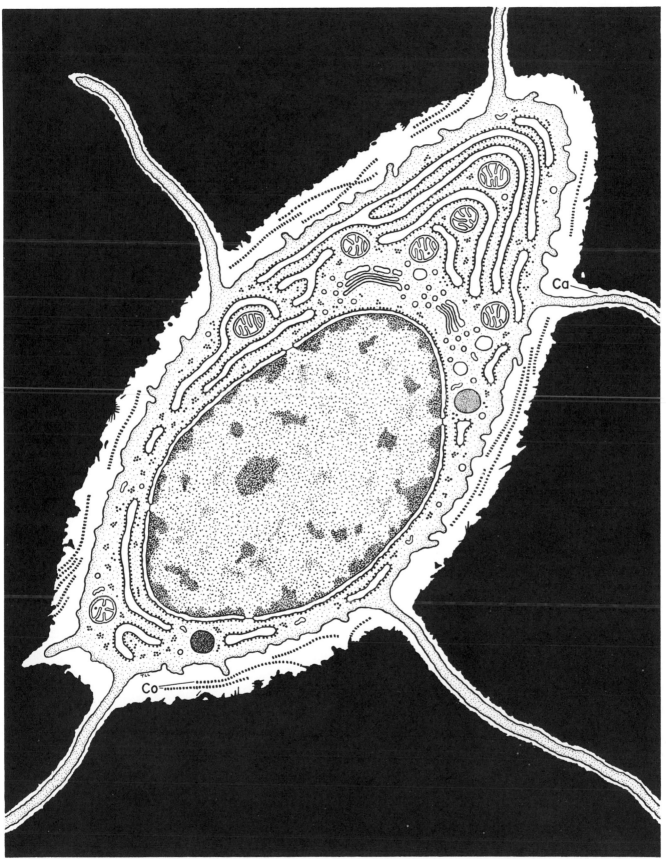

32—OSTEOCLAST

Osteoclasts are large multinucleated cells closely associated with areas of bone resorption. The cell surface in contact with resorbing bone (opaque material) is specialized as a ruffled border. This border is produced by complex infoldings of the plasma membrane that delimit irregular cytoplasmic arms. Small mineral crystals (Cry) liberated from the bone matrix are found at the edge of the ruffled border and also deep within the clefts. The cell surface in contact with nonresorbing bone (upper right of Plate) is unspecialized, and the underlying cytoplasm is devoid of organelles to form an ectoplasmic layer.

Organelles are concentrated in the perinuclear cytoplasm. Golgi complexes (G) are numerous and found adjacent to the nuclei. Granules (Gr) are abundant in the cytoplasm. One type of granule is small, round, oval, or elongated in profile, and composed of moderately dense material. These kinds of granules are especially numerous near the Golgi complexes. A second type is large and spherical, of low density, and composed of particulate or flocculent material. Some granules are intermediate in structure between these two basic types, and there appear to be transitional stages between the large granules and clear cytoplasmic vacuoles. The contents of the granules have not been identified, but osteoclasts contain acid phosphatase activity that could be contained in lysosomal granules. The granules found in these cells may also represent secretion of collagenolytic enzymes involved in degradation of bone matrix.

The endoplasmic reticulum of osteoclasts is sparse, consisting of a few short, rough-surfaced profiles. Ribosomes are found free in the cytoplasm, usually in clusters. Mitochondria are numerous and relatively large, with cristae extending across the organelle. A few irregularly shaped lipid droplets are found interspersed among the granules.

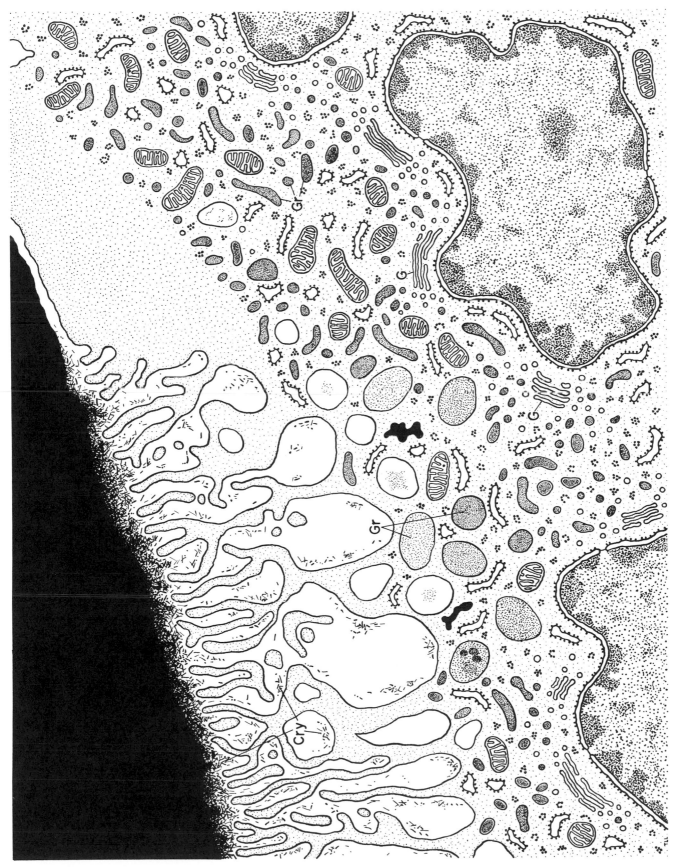

33—RETICULAR CELL

Reticular cells form the supporting framework of the lymphoid organs, including the lymph nodes, spleen, and bone marrow. They are highly irregular in shape with elongated branching cell processes, and they are closely associated with extracellular reticular fibers. Reticular cells are capable of several functions; this is reflected by a wide variability in cytoplasmic structure. As a major component of the reticuloendothelial system, they are phagocytic and may contain a number of lysosomes (Ly). Others have a relatively extensive rough-surfaced endoplasmic reticulum and appear to be engaged in protein synthesis. Reticular cells also trap antigens on their surfaces in lymph nodes. Lymphocytes in contact with the reticular cells are then induced to differentiate into antibody-producing plasma cells. Finally, some reticular cells have a sparse content of organelles and may correspond to undifferentiated reticular cells.

Reticular cells have a central, spherical nucleus with a nucleolus. One or more Golgi apparatuses are found near the nucleus. Mitochondria are abundant, and small vesicles can be identified in the cytoplasm. The endoplasmic reticulum is represented by ribosome-studded cisternae, which may be short, elongated and flattened, or markedly dilated (ER). Ribonucleoprotein particles are also found free in the hyaloplasm. In some regions, bands of dense material or bundles of filaments (Fl) occur beneath the plasma membrane.

These cells are responsible for the formation of reticular fibers. The fibers form delicate interlacing networks in the stroma of lymphoid tissues and around adipose cells, capillaries, muscle fibers, nerves, and the basal lamina of epithelia. Unlike collagenous fibers, the reticular fibers are argyrophilic, but they have the same structure as collagen when viewed with the electron microscope, and they are probably essentially the same.

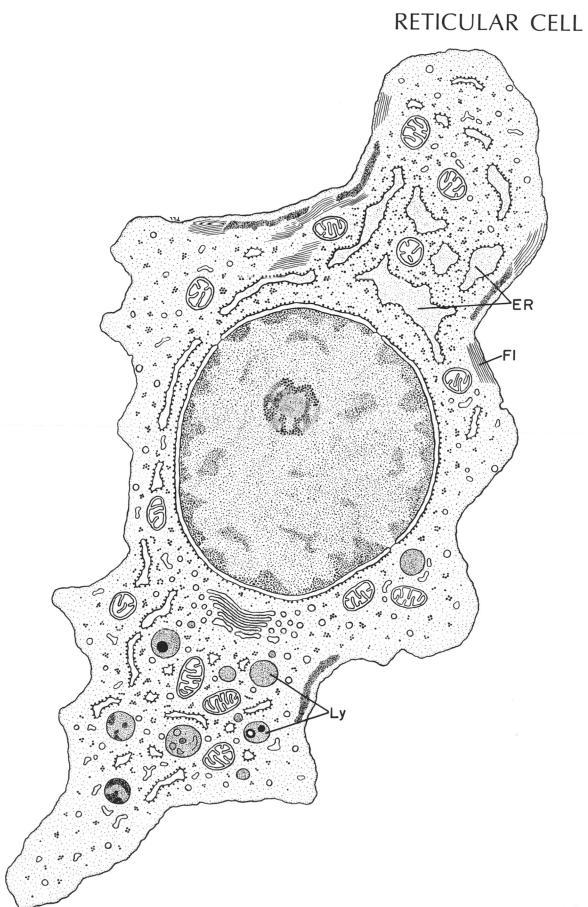

ER

FI

Ly

REFERENCES

Combs, J. W., 1966. Maturation of rat mast cells. An electron microscope study. J. Cell Biol., 31:563–576.

Han, S. S., J. K. Avery, and L. E. Hale, 1965. The fine structure of differentiating fibroblasts in the incisor pulp of the guinea pig. Anat. Rec., 153:187–210.

Napolitano, L., 1963. The differentiation of white adipose cells. J. Cell Biol., 18:663–680.

Napolitano, L., 1965. The fine structure of adipose tissues. *In* Handbook of Physiology. Sect. 5:109–124.

Revel, J.-P. and E. D. Hay, 1963. An autoradiographic and electron microscopic study of collagen synthesis in differentiating cartilage. Z. Zellforsch., 61:110–144.

Scott, B. L., 1967. The occurrence of specific cytoplasmic granules in the osteoclast. J. Ultrastruct. Res., 19:417–431.

Sutton, J. S. and L. Weiss, 1966. Transformation of monocytes in tissue culture into macrophages, epithelial cells, and multinucleated giant cells. An electron microscope study. J. Cell Biol., 28:303–332.

Thomson, J. F., D. A. Habeck, S. L. Nance, and K. L. Beetham, 1969. Ultrastructural and biochemical changes in brown fat in cold-exposed rats. J. Cell Biol., 41:312–334.

Wasserman, F. and J. A. Yaeger, 1965. Fine structure of the osteocyte capsule and of the wall of the lacunae in bone. Z. Zellforsch., 67:636–652.

Weiss, L., 1964. The white pulp of the spleen. The relationship of arterial vessels, reticulum and free cells in the periarterial lymphatic sheath. Bull. Johns Hopkins Hosp., 115:99–173.

MUSCULAR TISSUE

MUSCULAR TISSUE

34—SKELETAL MUSCLE: WHITE FIBER

Skeletal muscle fibers are elongated multinucleate cells containing parallel, banded myofibrils. The myofibrils are composed of smaller units, the myofilaments, whose relation to one another is responsible for the striated pattern of the fibrils. The striations of the fibrils are themselves in register, so that the entire muscle fiber has a band pattern that is visible with the light microscope. Thick myosin filaments 100 Å in diameter extend the length of the A band (A). The actin filaments are thinner, 50 Å in diameter. They extend from the Z line, constitute the I band (I), and overlap the thick filaments in the A band to the edge of the light H band (H), which is in the center of the A band. A denser M line (M), produced by cross connections between the thick filaments, bisects the H band. The sarcomere, considered to be the functional unit of contraction, is delimited by the dense Z lines (Z), which contain the protein tropomyosin.

According to the sliding filament hypothesis of contraction (Hanson and Huxley), the thin filaments slide past the thick filaments, shortening the lengths of the I band and the H band, thus reducing the length of the myofibril. Longitudinal displacement of the filaments relative to each other may be produced by the successive attachment of cross bridges on the myosin filaments to sites along the adjacent actin filaments. The energy for contraction is provided by the dephosphorylation of adenosine triphosphate (ATP) by adenosine triphosphatase (ATPase), which is localized in the cross bridges.

A system of membranous tubules is closely related to the myofibrils. Local differentiations in the membranous systems occur adjacent to particular bands in each sarcomere. In most mammalian muscle, near the level of the junction of the A and I bands are three circular profiles comprising the triad (Tr). The lateral profiles are larger than the central vesicles and are terminal dilatations of the smooth-surfaced endoplasmic reticulum or sarcoplasmic reticulum (SR). Small tubules of the sarcoplasmic reticulum run longitudinally between myofibrils. The small central vesicle of the triad represents a section through the transverse tubular system (TS), or T-system.

(Continued)

84

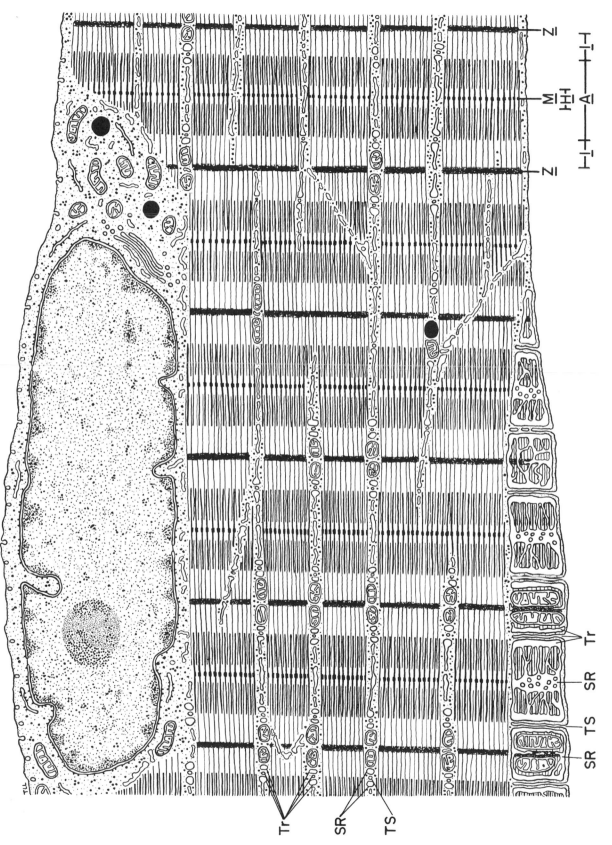

85

MUSCULAR TISSUE

34—SKELETAL MUSCLE: WHITE FIBER (Continued)

The relationship of the membranous structures to each other and to the fibril are apparent when viewed from their surface aspect, as in sections tangential to the fibril surface (lower left of Plate). In this situation, the transverse system is revealed as a tubule originating from and continuous with the sarcolemma and extending into the muscle fiber. On each side of the transverse tubule are the terminal cisternae of the sarcoplasmic reticulum that form transverse bands around the fibril. From the terminal dilated cisternae, longitudinal tubules run parallel to the surface of the fibril. Opposite the H band, the tubules run into a broad flat fenestrated cisterna that transversely encircles the fibril. The mitochondria opposite each I band are elongated and encircle the myofibril.

The membranous systems could be involved in the simultaneous activation of myofibrils throughout the muscle fiber following depolarization of the plasma membrane at the motor end-plate. The continuity of the transverse tubules with the sarcolemma indicates that these structures could be involved in inward spread of excitation from the sarcolemma. The sarcoplasmic reticulum, which is closely associated with the transverse system, then is thought to release calcium ions. The calcium ions seem to trigger contraction by activating ATPase, which causes hydrolysis of ATP with release of energy. Relaxation occurs when calcium is recaptured by the sarcoplasmic reticulum.

The nuclei are elongated and situated at the periphery of the muscle fiber. The nuclei contain one or two nucleoli, and chromatin is condensed adjacent to the inner side of the nuclear envelope. A small Golgi apparatus occurs near the end of the nucleus. The amount of sarcoplasm in relation to the myofibrils is small in white muscle fibers. A few small mitochondria occur in the sarcoplasm near the nucleus. In the interior of the fiber, mitochondria occur in pairs with each member of a pair opposite the I band. The mitochondria are filamentous and encircle the I band. In white muscle, the mitochondria are small and have sparse cristae; the matrix is of low density. A few lipid droplets occur in the sarcoplasm, but glycogen granules are abundant. Ribosomes are absent in the general sarcoplasm; however, they can be found beneath the motor end-plate.

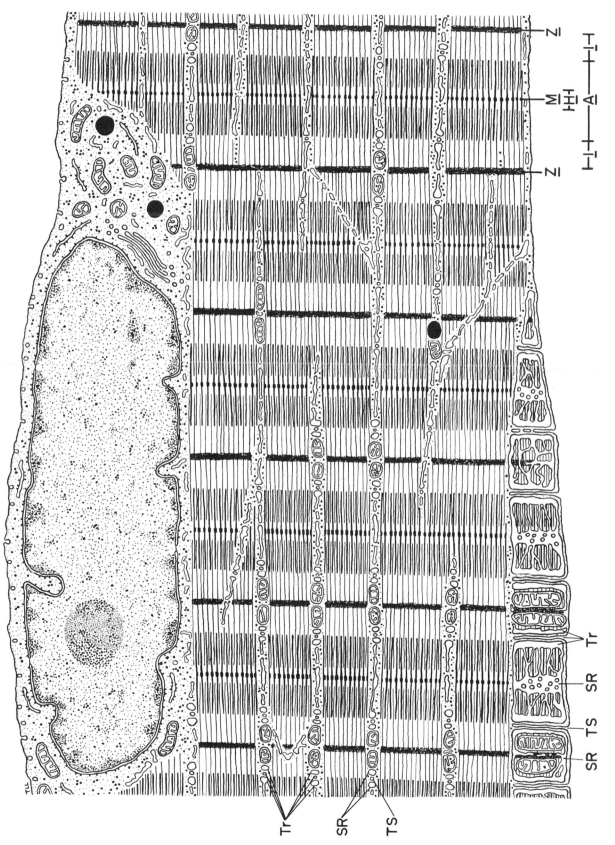

MUSCULAR TISSUE

35—SKELETAL MUSCLE: RED FIBER

Red muscle fibers have a smaller diameter than do white fibers, and the ratio of sarcoplasm to myofibrils is greater. A chief characteristic of this type of muscle is an abundance of mitochondria (M). Accumulations of large mitochondria are located beneath the sarcolemma and near the nucleus. Longitudinal chains of mitochondria occur along the myofibrils. At the level of the I band, mitochondria encircle the fibril. In contrast to white muscle, the mitochondria of red muscle are large, contain numerous cristae, and have a dense matrix. Lipid droplets (LD) are abundant, often occurring near the I band and in close relation to mitochondria. The structure of the transverse system (TS) and the sarcoplasmic reticulum (SR) is similar to that of white muscle. Triads occur near the A–I junctions. In red muscle, however, a network of tubules occurs over the H band, in contrast to the broad fenestrated cisternae that occur in white muscle in this region. Another difference is that the Z line is thicker in red muscle.

Although the functional differences between muscle fiber types have not been completely clarified, red muscle fibers appear to be associated with high metabolic activity and high frequency of contraction, whereas white muscle may be indicative of lower metabolic activity and a slower rate of contraction. The small size of red muscle fibers imparts a high surface–volume ratio that favors exchange of gases, ions, and metabolites. The abundance of mitochondria provides a readily available supply of energy, and the close association of lipid with mitochondria suggests extensive oxidative activity. The network of sarcoplasmic tubules over the H band provides a large membranous surface area that may be related to speed of contraction. An exact correlation between fiber type and speed of individual contraction (which must be distinguished from frequency of contraction) has not yet been made. Red fibers have been associated with muscles that have a slow speed of individual contraction, and white fibers with muscles that have a fast speed of individual contraction, but there appear to be exceptions to this rule. Some muscle fibers have characteristics intermediate between red and white fibers.

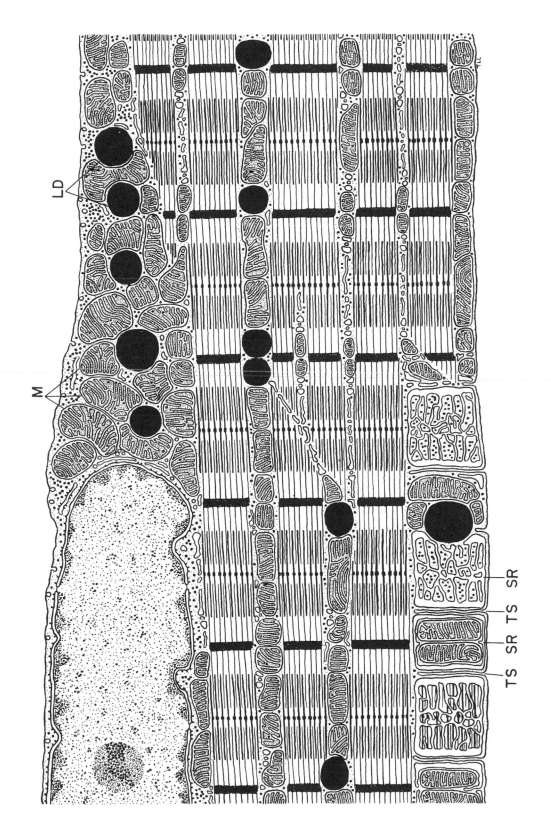

MUSCULAR TISSUE

36—SKELETAL MUSCLE: NEUROMUSCULAR JUNCTION

The motor axon terminates in a depression or gutter on the surface of the muscle fiber. Schwann cell (SC) cytoplasm covers the outer surface of the ending but is absent on the side facing the muscle cell. Neurofilaments and neurotubules course into the terminal axoplasm. Mitochondria and glycogen granules are present. Large numbers of small synaptic vesicles (SV) are clustered in the ending, especially in the axoplasm nearer the muscle fiber. A few of the vesicles are slightly larger and have a dense content.

The nerve ending is separated from the muscle surface by a space, the primary synaptic cleft. From the primary cleft, secondary synaptic clefts extend deep into the sarcoplasm. The ridges of cytoplasm between the secondary clefts are known as junctional folds (JF). The clefts are filled with amorphous material continuous with and similar to the basement membranes of the myofiber and the Schwann cell. This material is densest equidistant between muscle and axon surfaces.

The muscle cytoplasm is identified by myofibrils (lower left of Plate). A nucleus (lower right of Plate) may occur near the neuromuscular junction. The sarcoplasm underlying the junction contains mitochondria, some irregularly shaped membranous profiles, and some glycogen granules. Free ribosomes and a few short cisternae of rough-surfaced endoplasmic reticulum are also found in this region.

According to current information, the small synaptic vesicles of the nerve ending contain the neurotransmitter acetylcholine. Upon arrival of an action potential at the ending, acetylcholine is released from the vesicles, diffuses across the synaptic cleft, and brings about changes in the postsynaptic membrane's permeability to ions. The resultant depolarization is then propagated throughout the muscle fiber, leading to contraction (see Plate 34). Acetylcholine released from the vesicles is rapidly hydrolyzed by the enzyme acetylcholinesterase (AChE), which is localized predominantly on the postsynaptic membrane.

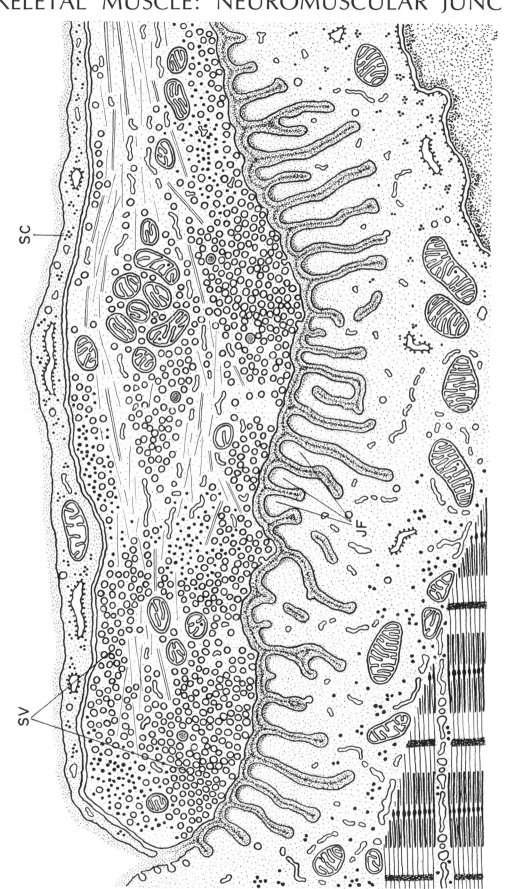

MUSCULAR TISSUE

37—SKELETAL MUSCLE: SATELLITE CELL

Satellite cells are small, flattened cells situated on the surface of muscle fibers. They are interposed between the plasma membrane and the basement lamina (BL) of the muscle fiber. The cells have a relatively uncomplicated structure. The nuclei are flattened, and the chromatin is more highly condensed than that in muscle nuclei. There is only a small amount of cytoplasm. Near the nucleus there are a small Golgi apparatus and a pair of centrioles. A few channels of rough-surfaced endoplasmic reticulum may be present; free ribosomes are more common. A few mitochondria are found. Pinocytotic vesicles are distributed along the surface facing the basement lamina, and some larger vacuoles occur in the cytoplasm. Satellite cells have been found to incorporate precursors of DNA and to undergo mitosis. It is thought that they may represent undifferentiated stem cells capable of giving rise to myoblasts in regenerating muscle.

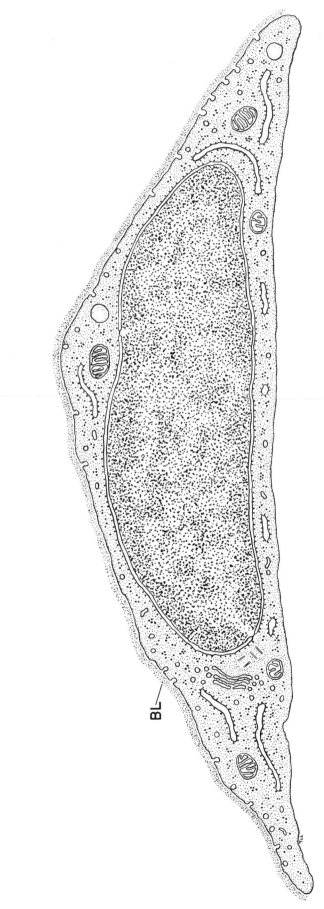

BL

93

MUSCULAR TISSUE

38—CARDIAC MUSCLE FIBER

Cardiac muscle and skeletal muscle are identically striated, but there are differences in other respects. The nucleus of cardiac muscle fibers is central in location and is surrounded by a zone of sarcoplasm rich in mitochondria. Bands of sarcoplasm separate regions of myofilaments and contain large mitochondria. Although the myofilaments appear to be separated into fibrils by the sarcoplasm, cross sections reveal that the area of myofilaments is confluent throughout the cell. The mitochondria have numerous angulated cristae and are often closely associated with lipid droplets. Glycogen granules are abundant in the sarcoplasm and occur between myofilaments, especially in the I band.

In contrast to skeletal muscle, the transverse tubules (TS) are located at the level of the Z line and are of larger diameter. The sarcoplasmic reticulum, on the other hand, is not as highly developed and consists of tubular elements extending roughly parallel to the filaments, with a few anastomoses and commissures. Small terminal dilations of the smooth cisternae may occur adjacent to the transverse tubules.

Cardiac muscle is composed of separate cellular units joined together by intercalated disks. The intercalated disks are specializations of the apposed surface plasma membranes at the transverse ends of the cells. The disk does not extend in a straight line across the width of the muscle fiber but, rather, in steps at different levels of the muscle. The disk contains regions corresponding to the junctional complexes occurring between epithelia (see Plate 78). Over the largest area of contact, the membranes are parallel and separated by a space of 200 Å. This specialization resembles the *zonula adherens* of epithelia, but because it covers an extensive area rather than a narrow belt, it is called *fascia adherens* (FA). Adjacent to these junctions is a zone of condensed cytoplasm into which the myofilaments insert. Desmosomes or *maculae adherentes* (MA) are usually located between these sites of insertion. The intercellular space is slightly wider in these regions, and a band of dense material is apposed to the inner surface of the plasma membrane. Small zones of membrane fusion occur along the sides of the undulations of the disk. These zones are small plaques and are termed *maculae occludentes* (MO). Desmosomes and large areas of membrane fusion, *fasciae occludentes* (FO), occur on the longitudinal surfaces where the cell processes overlap. The *fasciae occludentes* are believed to be sites of low electrical resistance that facilitate spread of excitation throughout the myocardium.

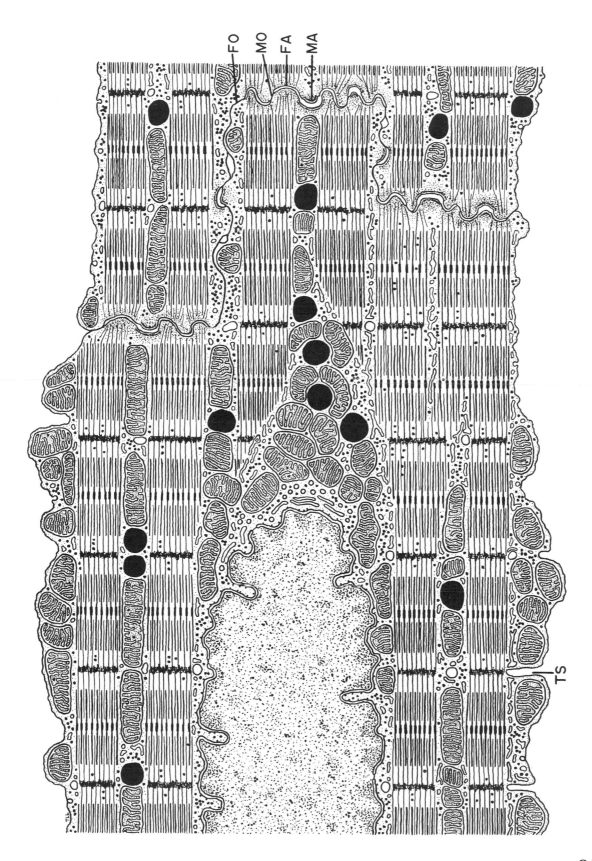

MUSCULAR TISSUE

39—PURKINJE FIBER

Purkinje fibers form the atrioventricular bundle (bundle of His), which is a part of the specialized conduction system of the heart. This system generates the stimulus for the heart beat and conducts it to the myocardium. Like the sinoatrial node and atrioventricular node, the atrioventricular bundle is composed of modified muscle fibers. The Purkinje fibers are large and have one or two nuclei with nucleoli. The hyaloplasm is of low density and appears to be empty, but it contains glycogen. Myofibrils occur but are narrow and are not composed of as many myofilaments as are most myofibrils. The fibrils are not aggregated into large bundles and are displaced toward the periphery of the cell. Mitochondria (M) are distributed throughout the fiber, and a few are unusual in structure, having only a single central crista. When sectioned transversely, they have the appearance of doughnuts. A few lysosomes or lipofuscin pigment granules are seen near the nucleus. The Purkinje fibers lack a transverse tubular system and thus have no triads. The sarcoplasmic reticulum (SR) is present but is not as extensive or as well organized as in the muscle fiber. Some cisternae are oriented in relation to the Z band, or closely parallel the cell surface. Pinocytotic vesicles are common along the surface as well. Purkinje fibers are joined to one another and to muscle fibers by junctional complexes with some of the components of the intercalated disk *(macula adherens, fascia adherens)*. In some places, close or tight junctions *(fascia occludens)* are found alone. These areas probably are low resistance couplings between adjacent cells. Thus, these junctions connect the Purkinje fibers with the myofibers, and these myofibers with other myofibers, consequently providing a continuous pathway for electrical conduction through the entire myocardium.

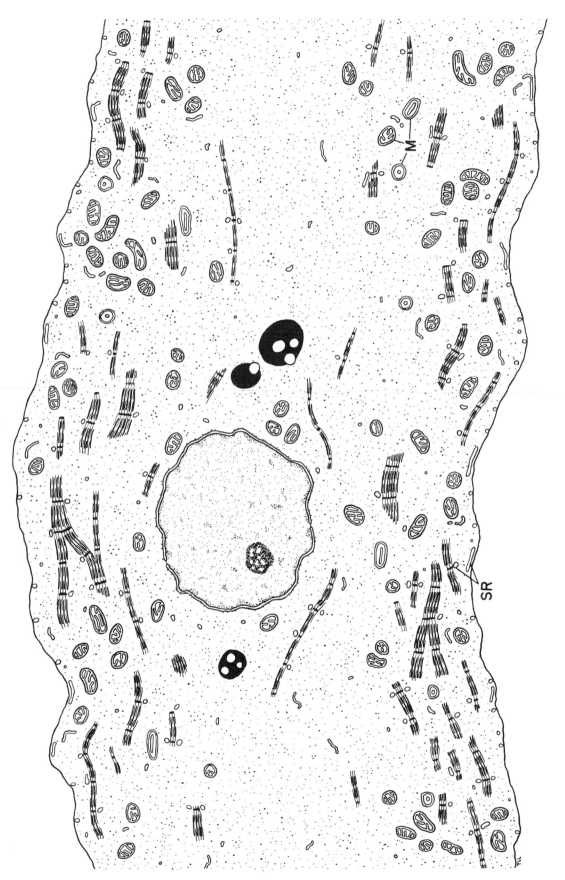

97

MUSCULAR TISSUE

40—SMOOTH MUSCLE CELL

Smooth muscle cells are fusiform or elongated in shape with a central, oval nucleus. Adjacent to the poles of the nucleus are regions of cytoplasm devoid of myofilaments. This portion of the cell contains a small Golgi apparatus, filamentous mitochondria, short cisternae of endoplasmic reticulum, and free ribosomes occurring in clusters. Longitudinal arrays of mitochondria and ribosomes also occur in other regions of the cell between filaments. The bulk of the cytoplasm is filled with myofilaments (Mf) arranged in the long axis of the cell and occurring in poorly defined bundles. Two types of filaments have been described: numerous thin filaments and a much smaller population of thicker filaments. The thicker filaments seem to be distributed randomly and to have no precise relation to the thin filaments.

Oval or elongated regions of hyaloplasmic density occur at intervals beneath the sarcolemma. Myofilaments seem to terminate in the subsarcolemmal densities. Densities in adjacent cells are located opposite one another, and pinocytotic vesicles are associated with the plasma membrane between densities. A coating of filamentous material invests the cell surfaces. A relatively large intercellular space occurs between the greater part of adjacent cell surfaces. However, in some regions the plasma membranes either fuse, obliterating the intercellular space, or are very closely apposed, forming close junctions. This junction, probably representing a *fascia occludens* (FO) and also called a tight junction or nexus, occurs where the cell surfaces are parallel, or where one cell is projected into another.

A major problem in explaining the mechanism of contraction of smooth muscle is that the structural counterpart of myosin has not been identified. The thin filaments that occur in large numbers are thought to correspond to actin. The paucity of thick filaments that would correspond to myosin has made it difficult to explain contraction on the basis of the filament interaction that occurs in striated muscle. One explanation is that myosin in relaxed muscle exists in a dispersed or soluble form, but during contraction it aggregates into a filamentous structure that interacts with actin to produce shortening. Thick filaments have been observed in contracted smooth muscle but are rare in relaxed smooth muscle.

98

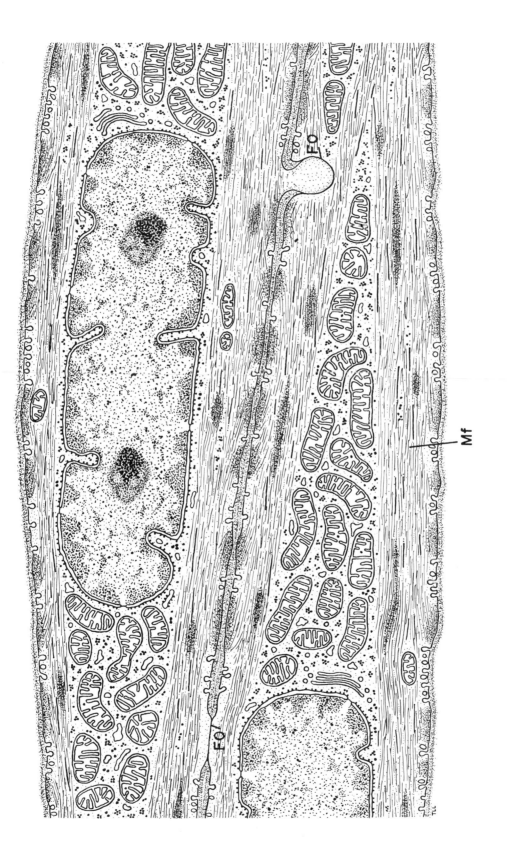

MUSCULAR TISSUE

41—MYOEPITHELIAL CELL

Myoepithelial cells are contractile cells found in many glands, including the sweat, salivary, lacrimal, mammary, and prostate glands. They are of ectodermal origin and lie on the epithelial side of the basement membranes (BL) of the glands and also beneath the secretory cells. The cells are usually elongated or branched and extend parallel to the long axis of the secretory segment or tubule. The upper surface of the cell is relatively smooth in contour and is associated with pinocytotic vesicles. The basal surface is more undulating and indented. On the cytoplasmic side of the basal plasmalemma, short plaques of moderately dense amorphous material immediately adjacent to the membrane alternate with regions of pinocytotic vesicles. Rarely, the cell is connected to an overlying secretory cell by a desmosome. Most myoepithelial cells are isolated from one another, but where they are apposed a close or tight junction may be found.

The oval nucleus is located apically in the central region of the cell. Most of the cell is occupied by masses of myofilaments (Mf) with organelles largely restricted to a region around the nucleus and to a thin rim beneath the apical surface. Oval to elongated mitochondria, glycogen granules, occasionally lipid and lipofuscin granules, and short cisternae of rough-surfaced endoplasmic reticulum are found in these areas. A small Golgi apparatus is located near the nucleus. Irregular tubules and vesicles of smooth endoplasmic reticulum are common. In places, some thin arms of cytoplasm penetrate the mass of filaments and usually contain smooth endoplasmic reticulum, glycogen, or elongated mitochondria.

The myofilaments run parallel to one another generally in the long axis of the cell. They are approximately 50 Å in diameter. Here and there among the filaments are elongated masses or bands of moderately dense amorphous material similar to those seen in smooth muscle. Many filaments appear to terminate on the dense plaques on the basal surface.

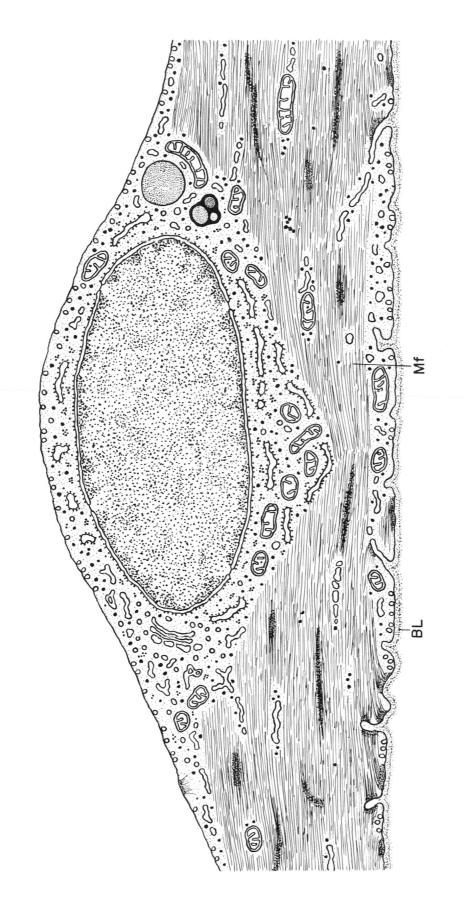

REFERENCES

Dewey, M. M. and L. Barr, 1964. A study of the structure and distribution of the nexus. J. Cell Biol., 23:553-586.

Ellis, R. A., 1965. Fine structure of the myoepithelium of the eccrine sweat glands of man. J. Cell Biol., 27:551-563.

Gauthier, G. F. and H. A. Padykula, 1966. Cytological studies of fiber types in skeletal muscle. A comparative study of mammalian diaphragm. J. Cell Biol., 28:333-354.

Ishikawa, H., 1966. Electron microscopic observations of satellite cells with special reference to the development of mammalian skeletal muscles. Z. Anat. Entwicklungsgesch., 125:43-63.

James, T. N. and L. Sherf, 1968. Ultrastructure of the human atrioventricular node. Circulation, 37:1049-1070.

Padykula, H. A. and G. F. Gauthier, 1970. The ultrastructure of the neuromuscular junction of mammalian red, white, and intermediate skeletal muscle fibers. J. Cell Biol., 46:27-41.

Porter, K. R. and G. E. Palade, 1957. Studies on the endoplasmic reticulum. III. Its form and distribution in striated muscle cells. J. Biophys. Biochem. Cytol., 3:269-300.

Sommer, J. R. and E. A. Johnson, 1968. Cardiac muscle. A comparative study of Purkinje fibers and ventricular fibers. J. Cell Biol., 36:497-526.

Tandler, B., 1965. Ultrastructure of the human submaxillary gland. III. Myoepithelium. Z. Zellforsch., 68:852-863.

VASCULAR TISSUE

VASCULAR TISSUE

42—CONTINUOUS CAPILLARY ENDOTHELIAL CELL

The walls of blood capillaries are composed of an inner layer of endothelium, a middle layer of basement membrane with associated pericytes, and an outer adventitia of cellular and extracellular connective tissue elements. Capillaries can be classified on the basis of specific morphological criteria that reflect different functional capacities. Capillaries of smooth, skeletal, and cardiac muscle, lung, connective tissue, and skin are similar. These capillaries have an uninterrupted layer of endothelium and are referred to as continuous capillaries.

One to three endothelial cells extend around the lumen of the capillary. The nucleus is oval and flattened and may cause a slight bulge of the inner cell contour into the lumen. The nuclear envelope contains pores and bears ribosomes on its outer surface. The nucleus contains a nucleolus and irregular masses of chromatin concentrated at the periphery. A pair of centrioles (Ce) and a Golgi apparatus (G) occur near the nucleus. The centrioles are situated within a mass of fine fibrillar material outside of which are located membranous lamellae and small vesicles of the Golgi complex. Multivesicular bodies or dense bodies are often found in the centrosphere region. The other cytoplasmic structures are concentrated around the nucleus and also occur in smaller numbers in the outer attenuated regions of cytoplasm. A few short cisternae of rough-surfaced endoplasmic reticulum and tubular profiles of smooth endoplasmic reticulum are randomly disposed in the cytoplasm. Free ribosomes are usually grouped in clusters, and some larger glycogen granules occur. Microtubules and filaments are also found, the latter either isolated or in small bundles. Mitochondria are small with few cristae.

A conspicuous feature of the endothelial cell is the presence of large numbers of vesicles (PV), 700 Å in diameter. The vesicles are found throughout the cytoplasm but are concentrated near both cell surfaces. Some vesicles are free in the cytoplasm, while others are immediately adjacent to the plasma membrane. A spectrum of images from apposition of vesicles to the plasma membrane, to fusion with the membrane so that

(Continued)

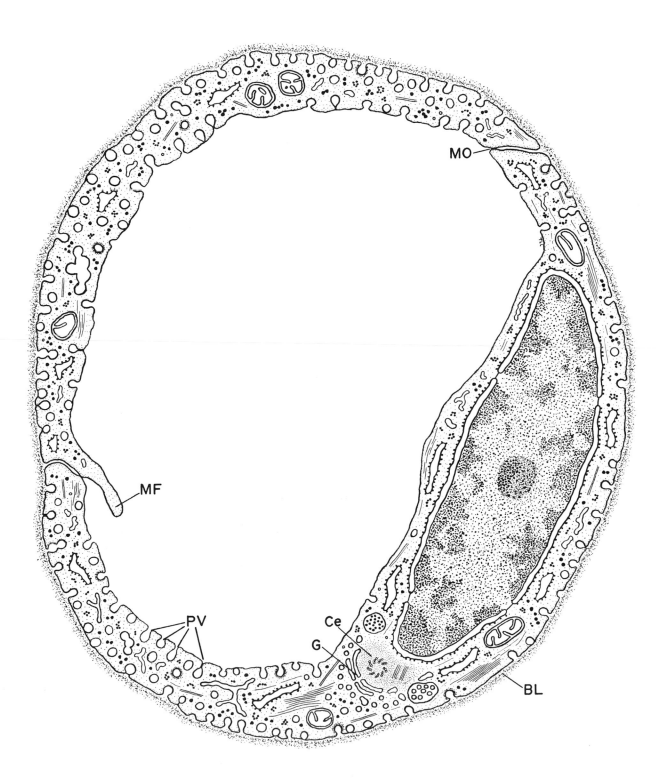

42—CONTINUOUS CAPILLARY ENDOTHELIAL CELL *(Continued)*

the vesicle opens to the surface have been described (Palade and Bruns). These modulations are thought to represent stages in the fusion of vesicles with the membrane and subsequent discharge of vesicle contents. Another series of images consists of flask-shaped vesicles connected to the membrane by a neck, elongation and pinching of the neck, and connection to the surface by a strand. These appearances may represent formation and pinching off of membrane invaginations (fission) to produce free vesicles. Some vesicles are fused to form chains that may be free in the cytoplasm or that open to the surface. A few vesicles have a coating of fine filamentous material on their outer surface (coated vesicles).

The margins of adjacent cells may bluntly abut end to end, overlap, or interdigitate. The margin of one or both cells may extend as a flap into the lumen, forming a marginal fold (MF). The membranes of adjoining cells are separated by a space of 100 to 200 Å along most of their adjacent lengths. However, near the lumen the membranes are in closer apposition, or they may be fused at a single point or over a large area. The tight junctions probably do not form a continuous belt but occupy smaller discrete areas (*maculae occludentes*, MO), because molecules, such as peroxidase, can pass through the intercellular clefts. A basement lamina (BL) consisting of filaments embedded in an amorphous matrix forms a continuous layer around the periphery of the endothelial cell.

The pinocytotic vesicles probably represent a mode of transport across the endothelial wall. Vesicles form on either cell front by invagination and pinching off from the plasmalemma, enclosing a quantum of extracellular material. They move across the endothelial cytoplasm, fuse with the plasmalemma of the opposite cell front, and discharge their contents. This type of transport may be related to the slow transcapillary exchange of high molecular weight molecules. A more rapid transit of smaller molecules may take place via the intercellular clefts (Karnovsky).

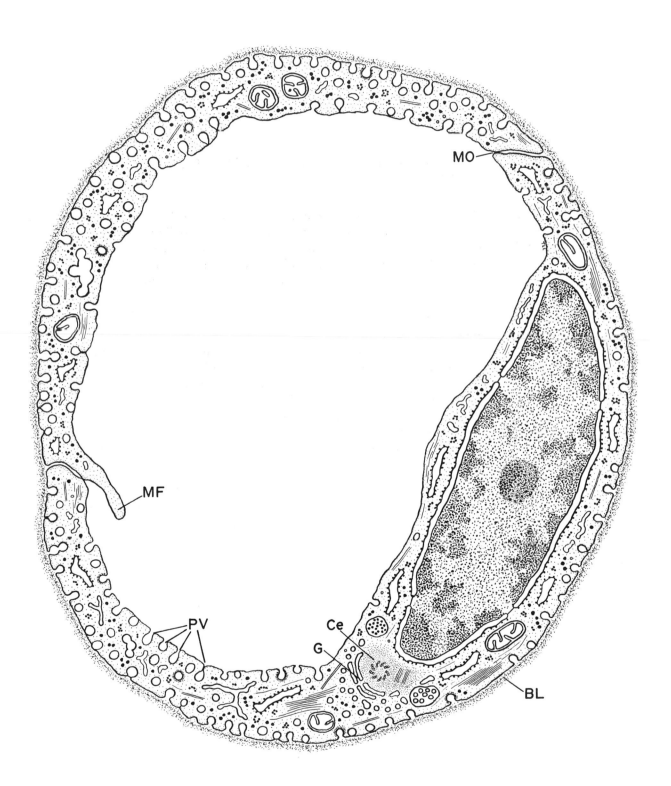

VASCULAR TISSUE

43—CONTINUOUS CAPILLARY ENDOTHELIAL CELL—CENTRAL NERVOUS SYSTEM

Capillaries of the central nervous system and retina are lined by a continuous layer of endothelial cells. No gaps or fenestrations occur in the endothelium. The capillaries are completely enveloped by a basement lamina, which splits to enclose the discontinuous layer of pericytes. Only a small extracellular space surrounds this type of capillary, because surrounding glial cells are closely applied to the basement membrane of the endothelium or pericyte.

These capillaries differ from other continuous capillaries in two important respects. First, the pinocytotic vesicles are not nearly as numerous, even though the general cytoplasmic structure is the same. The vesicles that are present open into both the luminal and perivascular surfaces. The second major difference is found in the junction formed by abutting endothelial cells. The plasma membranes of adjacent cells are fused in one or more places. The region of fusion often extends radially over much of the length of the junction. The regions of membrane fusion appear to form complete zonules or bands around the endothelial cells and thus represent *zonulae occludentes* (ZO). A small amount of dense material is sometimes applied against the inner surface of the plasma membrane along the tight junction.

In contrast to the process in continuous capillaries of muscle, intravenously injected tracers do not pass through the clefts between these endothelial cells. Transport across the endothelial cell via small vesicles likewise is limited. Thus, the tight junction and the lack of vesicular transport are morphological manifestations of the blood-brain barrier. This barrier prevents some substances, such as vital dyes, pigments, metals, and large molecules, which cross capillaries in other regions of the body, from entering the narrow intercellular spaces around neurons and glia.

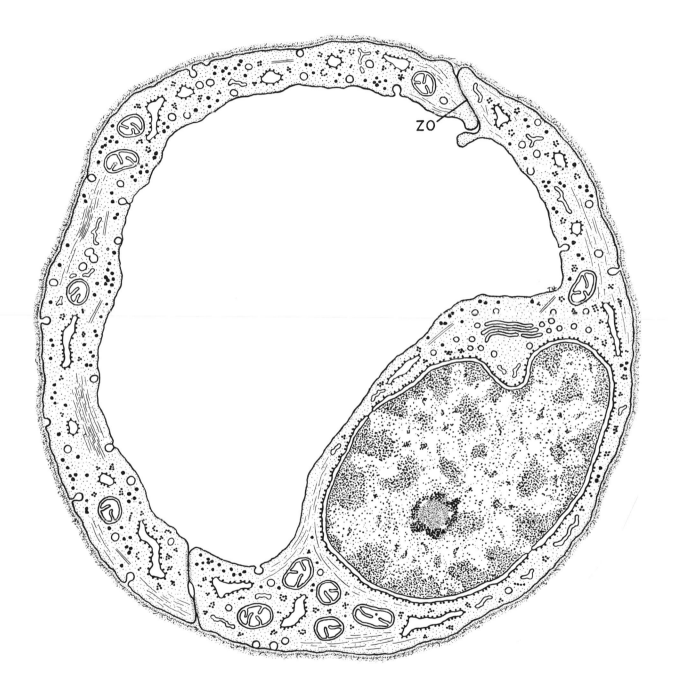

ZO

VASCULAR TISSUE

44—FENESTRATED CAPILLARY ENDOTHELIAL CELL

Fenestrated capillaries are found in the renal glomeruli, endocrine glands, intestinal villi, and exocrine pancreas. This type of capillary is characterized by the presence of pores or fenestrations (Fen) in the endothelial lining. The fenestrae occur in extremely attenuated regions of endothelium. The pores are 500 to 800 Å in diameter. They are bridged by a diaphragm that is composed of a single layer which has been interpreted as residual membrane proteins. A slight thickening or knob has been described in the center of the diaphragm. The endothelial wall is thicker between the thin regions containing fenestrations but is still generally thinner than the wall of continuous capillaries. The basement lamina surrounding the endothelial cell is continuous across the fenestrations. The glomerular capillaries (Plate 98) differ from other fenestrated capillaries in that, according to most investigators, there is no diaphragm over the fenestrae and the basement membrane is thicker.

In most other respects, the cytoplasmic structure of the fenestrated capillaries is similar to that of the continuous type. The crescent-shaped nucleus occurs on one side of the cell. A small Golgi apparatus, mitochondria, elements of endoplasmic reticulum, ribosomes, and glycogen granules are present in the cytoplasm. Pinocytotic vesicles occur but are not as numerous as in continuous capillaries. The fenestrae of these capillaries are considered to facilitate the transfer of material across the endothelium. In the case of the glomerular capillaries, transfer of fluid is extremely rapid.

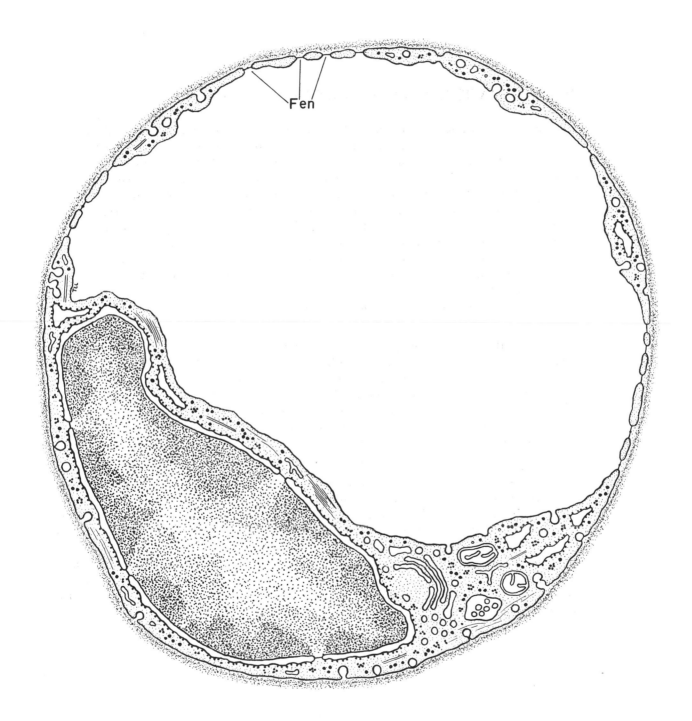

Fen

VASCULAR TISSUE

45 — LYMPHATIC CAPILLARY ENDOTHELIAL CELL

One or more flattened endothelial cells form the wall of the lymphatic capillary. The lumen is wider and more irregular in shape than those of blood capillaries. The irregularly shaped nucleus causes the cell to bulge into the lumen. The Golgi complex occurs near the nucleus and is composed of stacks of membranous lamellae and small vesicles. The endoplasmic reticulum is represented by short membranous cisternae with ribosomes on their outer surfaces. Rosettes of ribosomes also occur free in the cytoplasm. Mitochondria, which are spherical or oval, are uncomplicated in internal structure. The irregular luminal surface sends thin processes into the lumen. Pinocytotic vesicles occur in the cytoplasm and are associated with both surfaces. Some of the vesicles are coated. Both microtubules and thin filaments (Fl) are found in the cell cytoplasm. The filaments are especially abundant, and they are often aggregated into bundles. Some of the bundles course along the periphery of the cell beneath the plasmalemma.

Adjacent endothelial cells may overlap, interdigitate, or abut end to end. The junction formed at the margins has an intercellular space of about 200 Å over most of its length. Fusion of adjacent membranes (*macula occludens*) is sometimes seen near the lumen, but this does not always occur. Many of the junctions appear to be open along their entire length, indicating that materials could readily pass through them. In some places along the junction, dense material accumulates on the inner aspect of the plasma membrane to form small desmosomes.

The surface of the endothelial cell facing the tissue space has an irregular contour and sends projections or processes into the connective tissue space. The basal lamina (BL) is discontinuous, being interrupted by numerous filaments (Fl). These filaments often occur in bundles paralleling the lymphatic wall and seem to terminate on the outer surface of the plasma membrane. Filaments also emanate from the endothelial processes and at the margins of adjacent endothelial cells. Dense material accumulates in the cytoplasm adjacent to the inner aspect of the plasma membrane opposite to where the filaments insert. The filaments extend for long distances into the connective tissue, and it has been suggested that they could serve to anchor the lymphatics to the surrounding tissues.

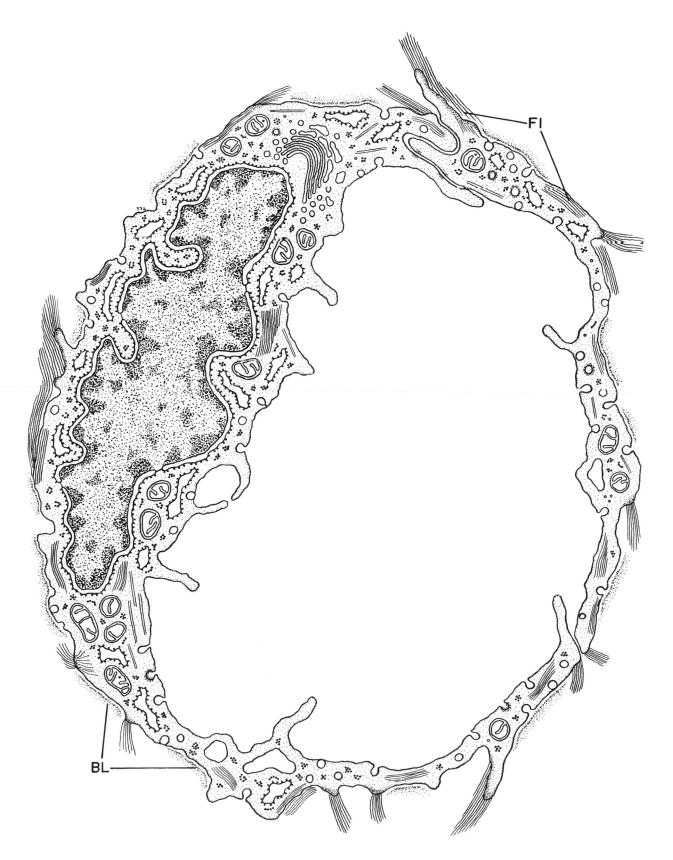

VASCULAR TISSUE

46—LITTORAL CELL

Littoral cells form the lining of the vascular sinusoids of the spleen, bone marrow, liver, and lymph nodes. They are components of the reticuloendothelial system and are phagocytic or potentially phagocytic. The sinuses formed by the littoral cells differ from vessels lined by endothelium. The cells are not always thin and attenuated but may be much thicker than those of endothelium. Gaps exist in the wall of the sinusoid. These may be large, and they could represent channels through the cell or spaces between adjacent cells. Fenestrations can also occur in thin regions of the lining, but these do not seem to be bridged by a diaphragm. There are also junctions between adjacent cells, and there may be small areas of membrane fusion. In contrast to endothelia, the basement lamina is discontinuous or entirely lacking in these sinuses.

The littoral cells are irregular in shape and, as already indicated, vary greatly in thickness. They are functionally similar to other components of the reticuloendothelial system (reticular cells, Plate 33; pericytes, Plate 47) in that they are capable of phagocytosis. The cytoplasm is rich in mitochondria and contains cisternae of rough-surfaced endoplasmic reticulum, lysosomes, free ribosomes, and a Golgi complex. Pinocytotic vesicles are abundant in littoral cells, especially near the cell surfaces. A band of dense material (lower right of Plate) is often seen in these cells, usually near the surface of the cell facing the subsinusoidal space. The space contains the collagenous fibers and ground substance comprising the extracellular reticulum.

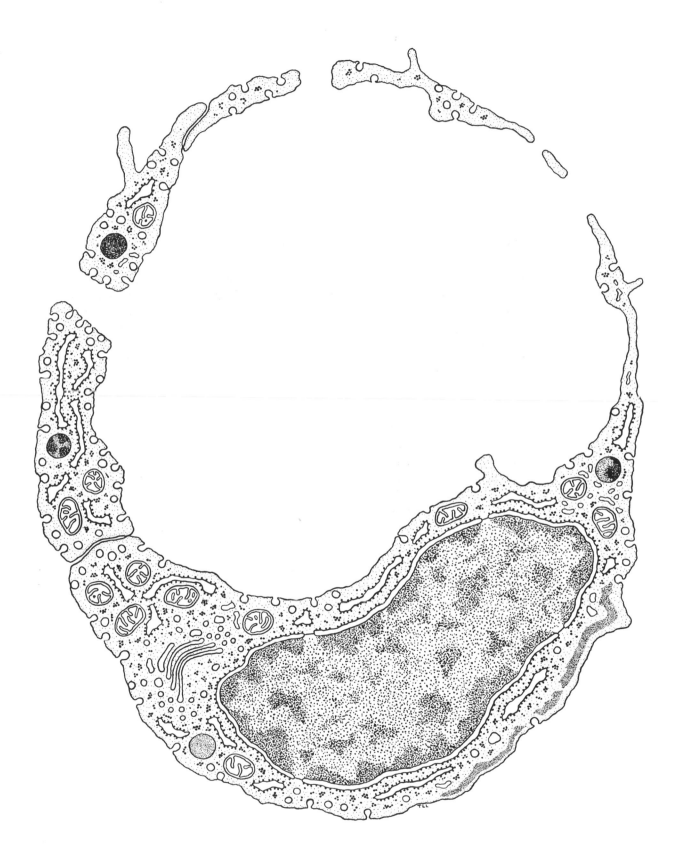

47 — PERICYTE

Pericytes (adventitial cells, Rouget cells) occur outside the endothelial lining of arterioles, capillaries, and sinuses. In capillaries (CL indicates the capillary lumen), they are located between leaflets of the basement lamina. The cells are polymorphous, some having long cytoplasmic processes extending from the perikaryon along or encircling the vessel. The nucleus causes the cell body to protrude toward the perivascular space. A Golgi apparatus occurs near the nucleus. Elements of both smooth- and rough-surfaced endoplasmic reticulum are present. The latter occurs as flat, elongated cisternae. Free ribosomes are found as well as larger dense granules representing glycogen. Mitochondria are numerous and relatively large. Lysosomes occur throughout the cytoplasm. Pinocytotic vesicles are associated with the cell surface.

Filaments (Fl) are abundant in pericytes and often occur in bundles. An especially prominent bundle of filaments courses beneath the surface of the cell facing the endothelium. Pinocytotic vesicles are absent along this cell front. The bundle of fibrillar material, or sole, extends between small foot processes. The foot processes penetrate the basement membrane and make contact with the endothelium. A small area of membrane fusion, *macula occludens* (MO), may occur in these areas.

The function of pericytes is poorly understood. It has been suggested that they are contractile and can play a role in regulation of the size of the vessel lumen. The tight junctions between pericytes and endothelial cells are indicative of low resistance coupling or ion transfer. Pericytes are capable of phagocytosing certain materials and are considered a component of the reticuloendothelial system.

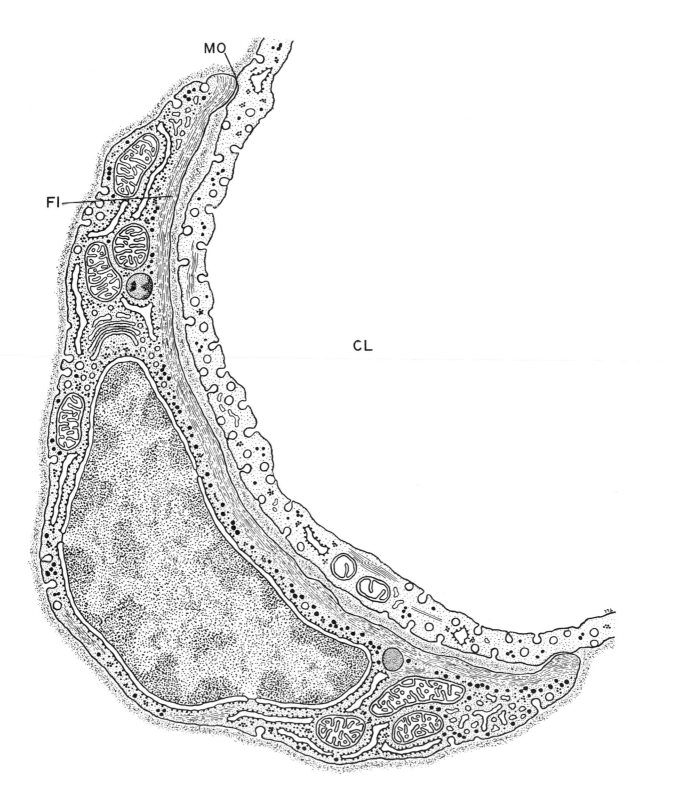

MO

FI

CL

REFERENCES

Bruns, R. R. and G. E. Palade, 1968. Studies on blood capillaries. I. General organization of blood capillaries in muscle. J. Cell Biol., 37:244-276.

Bruns, R. R. and G. E. Palade, 1968. Studies on blood capillaries. II. Transport of ferritin molecules across the wall of muscle capillaries. J. Cell Biol., 37:277-299.

Karnovsky, M. J., 1967. The ultrastructural basis of capillary permeability studied with peroxidase as a tracer. J. Cell Biol., 35:213-236.

Leak, L. V. and J. F. Burke, 1966. Fine structure of the lymphatic capillary and the adjoining con-nective tissue area. Amer. J. Anat., 118:785-810.

Palade, G. E. and R. R. Bruns, 1968. Structural modulations of plasmalemmal vesicles. J. Cell Biol., 37:633-649.

Reese, T. S. and M. J. Karnovsky, 1967. Fine structural localization of a blood-brain barrier to exogenous peroxidase. J. Cell Biol., 34:207-218.

Rhodin, J. A. G., 1965. Ultrastructure and function of liver sinusoids. Proc. IVth Intern. Symp. Reticuloendothelial System, 108-123.

Wolff, J., 1963. Beiträge zur Ultrastruktur der Kapillaren in der normalen Grosshirnrinde. Z. Zellforsch., 60:409-431.

SKIN AND OTHER EPITHELIA

48 — EPIDERMIS: CELL OF THE STRATUM GERMINATIVUM

The cells of the stratum germinativum or stratum basale are in contact with the dermis. These cells divide and differentiate giving rise to cells of the upper layers of the epidermis, eventually replacing the cells of the stratum corneum lost by desquamation. The germinative cells are polygonal in shape and have a central nucleus. The nucleus contains one or more nucleoli, and a thin layer of condensed chromatin is applied to the inner aspect of the nuclear envelope. The cytoplasm contains large numbers of tonofilaments (Tf), many of which are organized in bundles or tonofibrils. A few short cisternae of rough-surfaced endoplasmic reticulum are found, while clusters of free ribosomes are abundant. A number of small mitochondria and melanin granules or melanosomes (Mel) occur in the cytoplasm. The melanosomes are derived from the melanocytes (Plate 53) and may occur singly or in clusters of varying size enveloped by a membrane. The melanin granules are most abundant in cells of the lower layers of the epidermis.

Both the basal and lateral surfaces are irregular with low folds and undulations. Desmosomes or *maculae adherentes* (MA) are common between adjacent cells. In these regions, the intercellular space is about 240 Å across and is occupied by moderately dense material. A central dense line running parallel to the plasma membranes, sometimes flanked by additional lines, may occur in this dense material. A plaque of dense cytoplasmic material is found against the inner aspect of the plasma membrane. Some of the cytoplasmic tonofilaments converge upon the dense material. At the basal surface of the cell that overlies the dermis but is not in immediate contact with other cells, half desmosomes or hemidesmosomes are found.

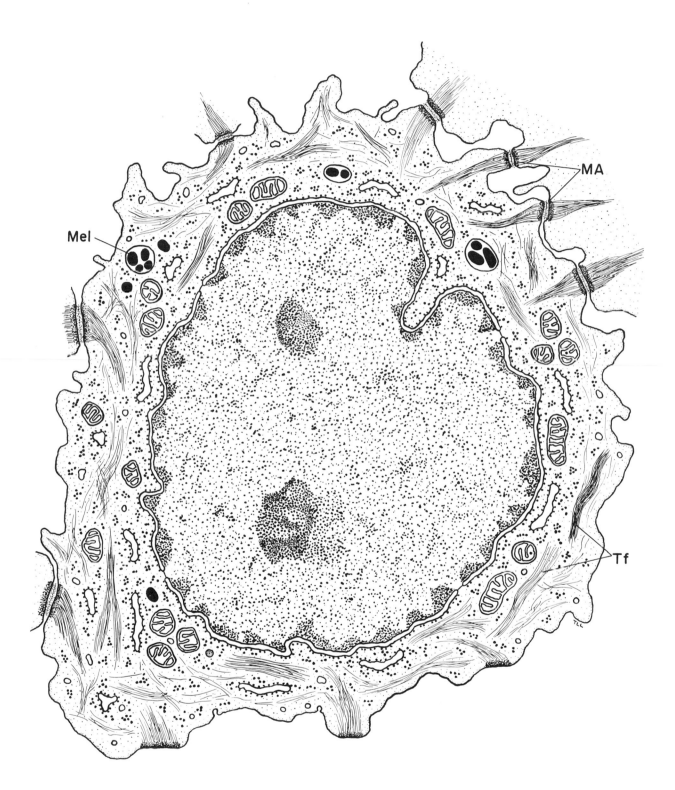

49—EPIDERMIS: CELL OF THE STRATUM SPINOSUM

The stratum spinosum overlies the basal layer of germinative cells. Together, the stratum basale or germinativum, and the stratum spinosum comprise the stratum malpighii. The cells of the stratum spinosum represent the first stage in differentiation of the basal cells in the process of keratinization. In the upper layers of the stratum spinosum, the cells are somewhat flattened and have a central nucleus containing a nucleolus. The cytoplasmic tonofilaments (Tf) of these cells are better organized into bundles or tonofibrils that tend to run parallel to the surface of the skin. The proteins of the tonofilaments are probably synthesized by the ribosomes and represent the first stage in the formation of the fibrous protein keratin. Small granules (Gr) spherical to ovoid in shape are also found in the uppermost layers. These structures appear to arise in the vicinity of the Golgi apparatus and have a dense, lamellar internal structure. It has been suggested that these granules fuse with the plasma membrane and liberate their contents, which coat the cell surface with a protective layer. For this reason, they have been termed membrane-coating granules. The surfaces of the cells are highly irregular with many ridges and cytoplasmic projections or spines (Sp). These processes contact those of adjacent cells, and it is in these regions that a desmosome (*macula adherens*, MA) is found. Mitochondria are distributed around the nucleus. Cytoplasmic ribosomes are not as abundant here as in the cells of the basal layer.

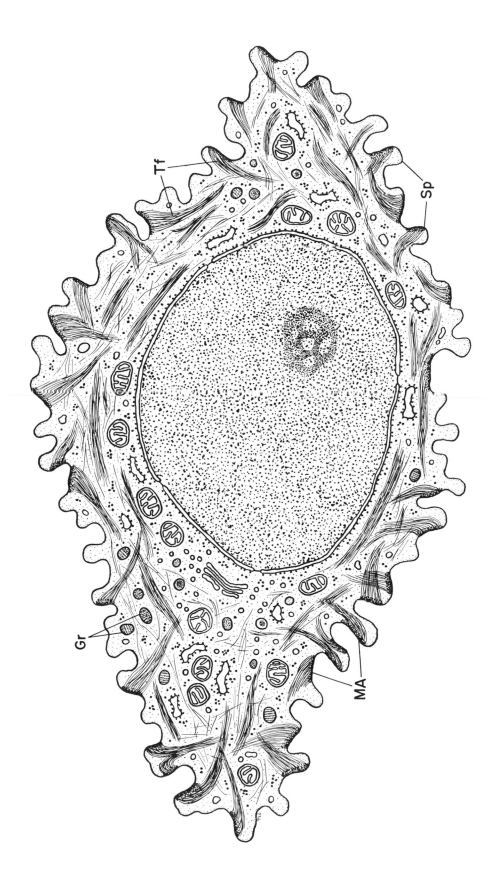

SKIN AND OTHER EPITHELIA

50—EPIDERMIS: CELL OF THE STRATUM GRANULOSUM

The cells of the stratum spinosum give rise to the cells of the stratum granulosum, which are arranged in one to four layers. The cells of the stratum granulosum are more flattened than those of the stratum spinosum, and they contain keratohyalin granules (KG). These consist of irregular, angular masses of dense material situated in association with the bundles of tonofilaments (Tf). It is thought that the dense material is synthesized by these cells and then deposited among the pre-existing tonofibrils. A few mitochondria are found, and ribosomes occur in the cytoplasm between keratohyalin granules. The nucleus is flattened and may lack the nucleolus. The surface projections are not as numerous or prominent as those in cells of the stratum spinosum, but desmosomes persist. When the keratohyalin granules reach maximum size, the cell loses its nucleus and cytoplasmic organelles and, at the same time, its capacity for protein synthesis.

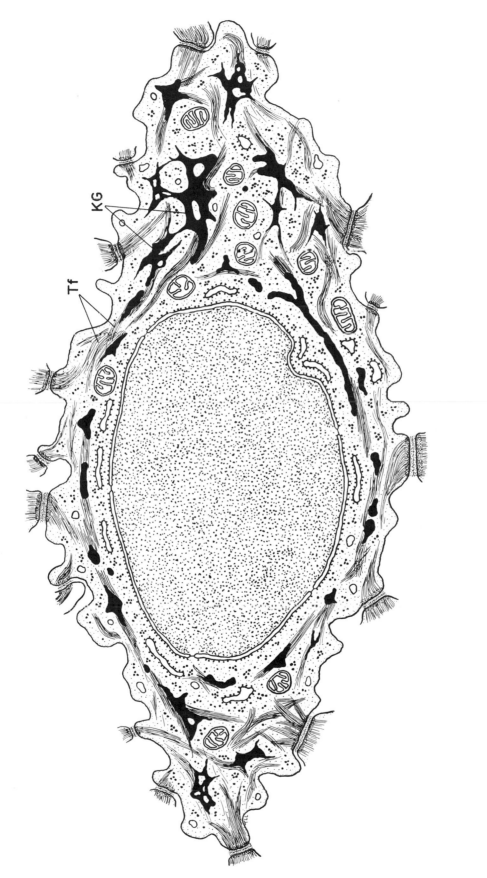

51 — EPIDERMIS: CELL OF THE STRATUM CORNEUM

The cells of the stratum granulosum flatten and lose their nuclei and organelles, giving rise to the cells of the stratum corneum. In regions such as the palms and soles where the epidermis is thick, the stratum lucidum intervenes between the stratum granulosum and the stratum corneum, and the stratum corneum is especially thick. The cells of the stratum corneum are broad flat scales. The cytoplasm of these cells contains no organelles but is filled with tightly packed, fine filaments embedded in a dense matrix. The filaments are composed of the protein keratin and may represent the tonofilaments formed in the lower layers, while the dense interfibrillar matrix may be the keratohyalin synthesized in the stratum granulosum. The surface of the cells may contain broad ridges and angulations, but the abundant projections and spines of the lower layers have disappeared. Desmosomes have also undergone considerable change in structure and are represented only by a dense plaque at the site of the original intercellular cleft. The plaque is separated from the thickened cell membranes by a thin cleft, and it eventually degenerates and disappears. These cells are the end product of the process of keratinization. In the final stages of keratinization, sulfhydryl groups of proteins are oxidized to disulfides, which are thought to be responsible for the great resistance of the stratum corneum to many substances including enzymes such as trypsin, dilute acids or alkali, and organic solvents. The superficial cells of the stratum corneum are constantly desquamated, and then replaced by differentiating cells from the deeper levels.

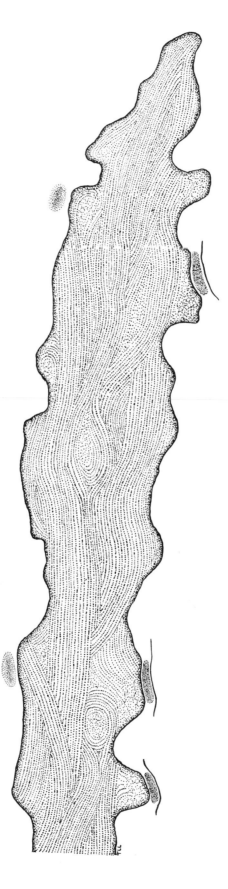

52—EPIDERMIS: LANGERHANS CELL

Langerhans cells are found singly in the lower and middle layers of the epidermis. They are irregular in shape, with processes extending between the keratinocytes. The central nucleus is lobulated, and the chromatin is condensed peripherally. The Golgi complex is well-developed and displays stacks of membranous lamellae and large numbers of small vesicles. The rough-surfaced endoplasmic reticulum is represented by scattered, isolated, short cisternae. Free ribosomes are common. Unlike other epithelial cells, Langerhans cells have abundant smooth-surfaced endoplasmic reticulum. Mitochondria, round to elongated in profile, are common, and a few lysosomes occur. Melanosomes and cytoplasmic filaments are few in number, and desmosomes are absent.

The characterizing feature of these cells is the Langerhans cell granule. The granules (Gr) are most abundant in the Golgi zone but are distributed throughout the cytoplasm. They display a variety of profiles in two dimensions. Commonly, they appear as rod-shaped structures composed of a limiting membrane enclosing a row of 90-Å particles. The rod-shaped profiles may be continuous with a dilated vesicle at one end, producing a configuration that resembles a tennis racket. Less commonly, the vesicle occurs in the center of the rod. The different profiles can be accounted for by a disk-to cup-shaped structure with a vesicular portion at one edge. Neither the origin nor function of the Langerhans cell is known. Most of the available evidence indicates that it is not a derivative of the melanocyte but is a distinct cell type.

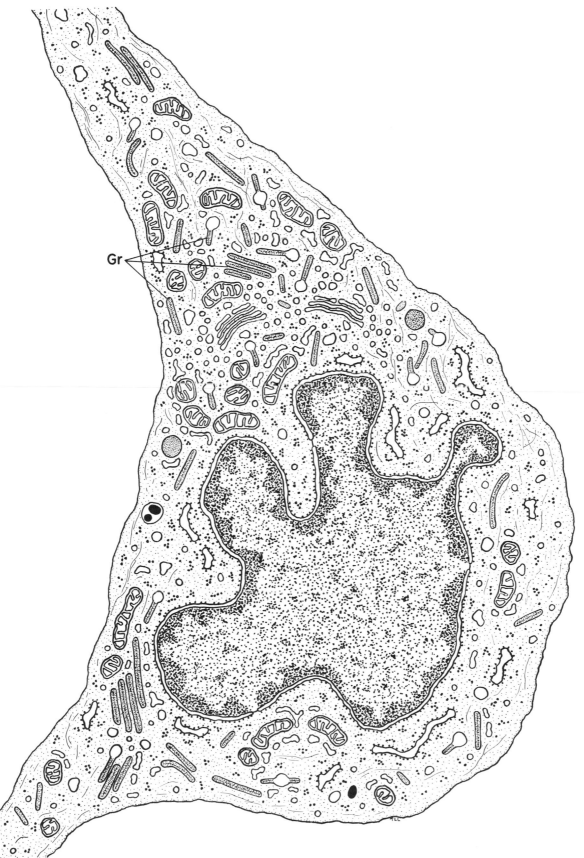

Gr

53—EPIDERMIS: MELANOCYTE

The epidermal melanocytes are situated in the basal layer of the epidermis in a ratio of about one pigment cell to four to twelve epidermal cells. The cells have an irregular surface but are not connected to other cells by desmosomes. Thin dendritic processes extend upward between the epidermal cells of the malpighian layer. The melanocytes have a round basal nucleus, which may be indented. A well-developed Golgi apparatus occurs near the nucleus. Mitochondria are relatively abundant. The cells have a few cytoplasmic filaments and cisternae of rough-surfaced endoplasmic reticulum as well as cytoplasmic ribosomes.

Melanocytes, which contain the enzyme tyrosinase, are responsible for synthesizing the black pigment melanin from tyrosine. The melanin is deposited on secretory granules known as melanosomes (Mel). These oval bodies are about 0.6 to 0.7 μ long and have a complicated interior structure of dense concentric lamellae. The material composing the lamellae shows periodicity. The interior structure of the melanosome is obscured with the accumulation of melanin. The melanin granules are transferred to the malpighian cells. The malpighian cells may accumulate large amounts of pigment, but the actively synthesizing and secreting melanocyte itself retains little.

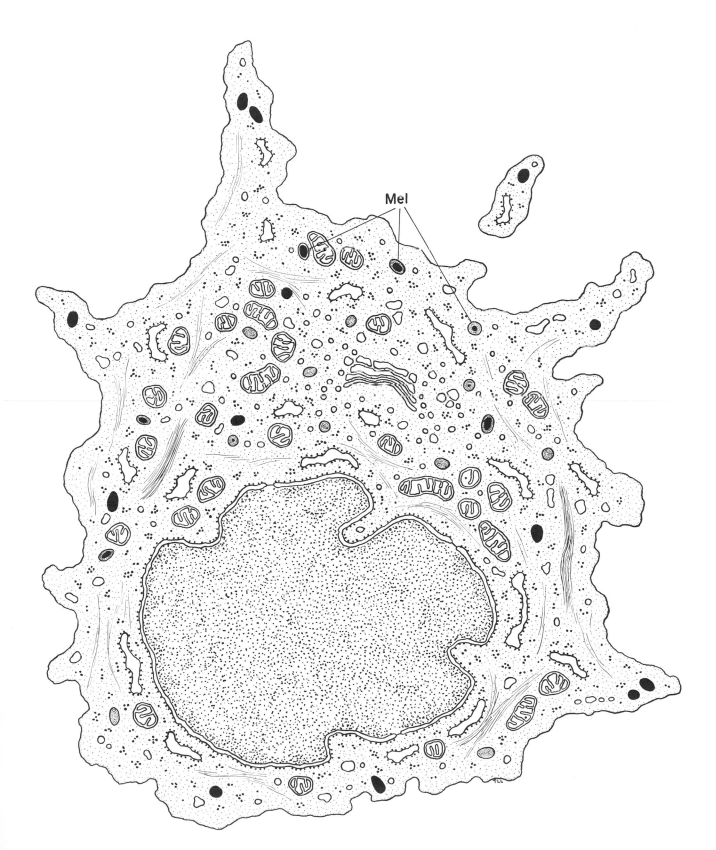

Mel

54—ECCRINE SWEAT GLAND: DARK CELL

The eccrine sweat gland is the ordinary sweat gland found over the entire surface of the skin with the exception of the lips, glans penis, and nail bed. The terminal or secretory portion of the long, tubular gland is coiled and lies in the dermis. Two cell types have been identified in the secretory segment: the dark cells and the clear cells. Myoepithelial cells (Plate 41) are arrayed along the periphery of the tubule.

The dark cells of the eccrine sweat gland border the main glandular lumen. These cells are columnar with a basal nucleus and have the usual complement of organelles found in secretory cells. Free ribosomes and elongated cisternae of rough-surfaced endoplasmic reticulum fill the base of the cell. The Golgi apparatus, supranuclear in position, is well-developed and has many small secretory granules associated with it. Large secretory granules fill the apical regions of the cell. The granules are thought to be rich in mucopolysaccharides and mucoproteins. Elongated mitochondria and a few glycogen granules are interspersed among these elements. Some stubby microvilli are found on the luminal surface, and there are some lateral infoldings, though they are not nearly as complicated as those of the clear cell. Other cells found in the eccrine glands have characteristics intermediate between those of clear cells and dark cells.

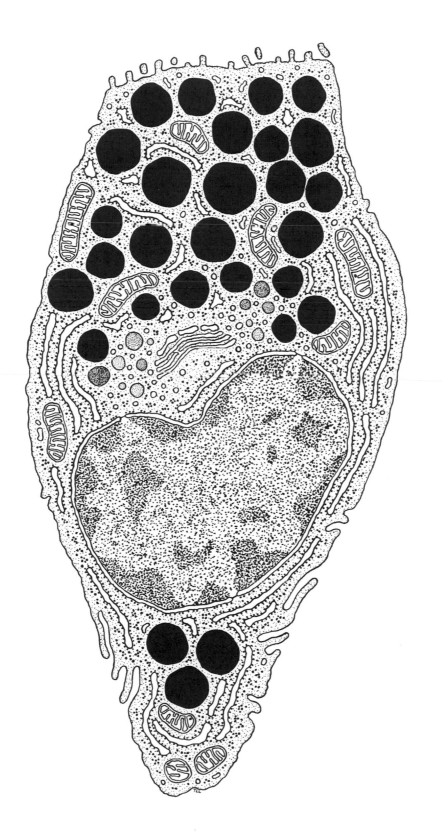

SKIN AND OTHER EPITHELIA

55—ECCRINE SWEAT GLAND: CLEAR CELL

The secretory portions of the eccrine sweat glands consist of clear cells, dark cells, and myoepithelial cells. The clear cells rest on the basement membrane or on myoepithelial cells. They rarely reach the lumen of the gland but border on small intercellular canaliculi (Ca). The cells are round or polygonal in shape and have a central nucleus. Their most conspicuous cytoplasmic feature is large accumulations of glycogen (Gly) that fill vast stretches of cytoplasm. Large oval mitochondria are also common and have numerous cristae that often extend across the organelle. A Golgi apparatus occurs near the nucleus, and associated with it are a few small dense granules. There are not many cisternae of endoplasmic reticulum, but some free ribosomes occur. Adjacent clear cells have complex interdigitating processes. These are thin and unusually long, and they often arise from wider cytoplasmic extensions that contain mitochondria. Some short microvilli extend into the lumen of the canaliculus. The clear cell bears some resemblance to the cells of the avian salt gland, and it has been suggested that the clear cell could play a role in salt excretion. Sweat is an aqueous product containing several solutes including chlorides, phosphates, urea, ammonia, and uric acid.

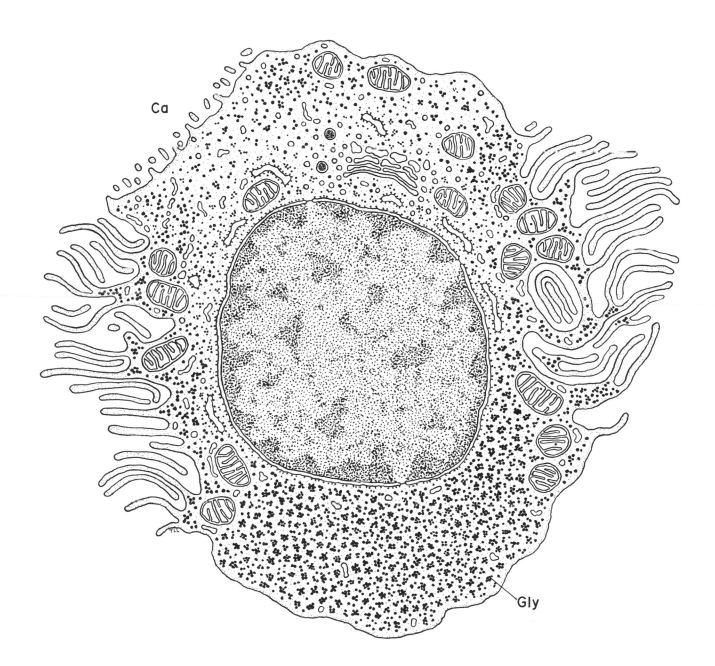

56—APOCRINE SWEAT GLAND CELL

Apocrine sweat glands are found in the axilla, pubic region, and around the anus. They differ from eccrine glands in being larger, being connected with hair follicles, and producing a more viscous secretion. The ceruminous glands of the external auditory meatus and the glands of Moll in the margin of the eyelid are specialized apocrine sweat glands. Cells of the secretory portion of the coiled tubular apocrine sweat glands are columnar with a central or basal nucleus. The apical portion of the cell bulges into the lumen forming a cytoplasmic cap. Some short microvilli are scattered along the apical surface. Laterally and especially basally, the plasma membranes are interdigitated.

Ribosomes and cisternae of rough-surfaced endoplasmic reticulum are abundant in the base of the cell. A prominent Golgi apparatus overlies the nucleus. Secretory granules (SG) composed of a low density flocculent material arise from the Golgi complex and are progressively larger toward the apex of the cell. The contents of the granules, which include a mucopolysaccharide, are liberated by fusion with the apical plasma membrane. This type of secretion is of the merocrine type, rather than of the apocrine type (in which part of the apical cytoplasm is lost). In the human, secretory granules are few in number, while other granules (Gr) containing sulfhydryl and disulfide groups and possibly keratin are common. These complex structures contain bundles of fine filaments intermingled among variable accumulations of electron-opaque material that is probably lipid in nature. The latter material may accumulate and obliterate other structures within the granule. Sometimes the central region of the granule is of low density and devoid of structure. Some smaller dense granules may also be found. Mitochondria (M) in the human are large and contain a few long cristae extending into an unusually dense mitochondrial matrix. A few fine filaments run in the cytoplasm between organelles. The apical cytoplasmic cap contains fewer large organelles than does the cell proper.

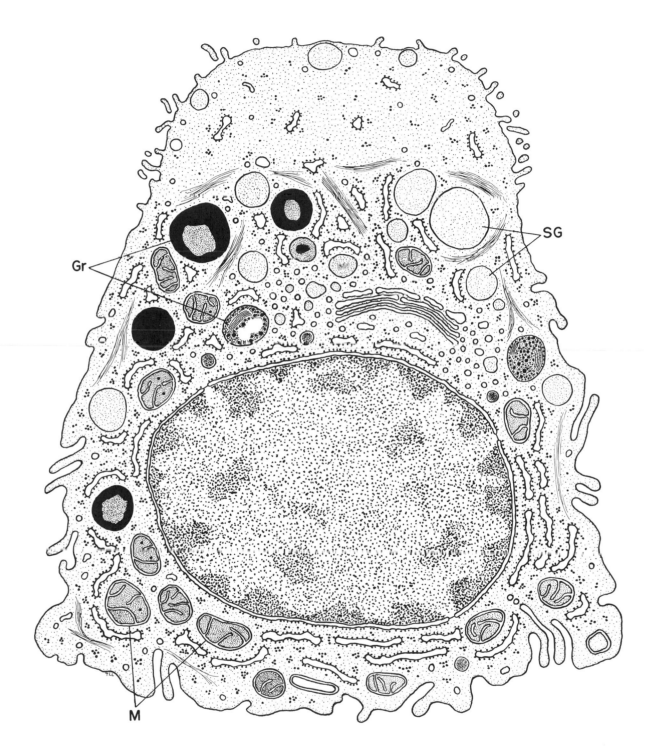

57 — SEBACEOUS GLAND CELL

Sebaceous glands are distributed over the entire skin surface except in the palms and soles. They are simple, branched alveolar glands that in most areas empty into the necks of hair follicles. The meibomian glands of the eyelids are specialized sebaceous glands. Cells of the sebaceous gland are large polyhedral cells filled with lipid droplets (LD). Secretion is of the holocrine type, in which the cells break down and add their content to the gland's oily secretion product called sebum. In actively synthesizing cells, lipid droplets are not as densely packed and are separated by greater amounts of cytoplasm. The lipid droplets are often composed of materials of different densities: moderately dense homogeneous material that tends to be distributed at the periphery of the droplet, and lucent material that is usually centrally located. The content of the lipid droplets includes cholesterol, cholesterol esters, and phospholipids. The lipid droplets are bordered by short cisternae and vesicles of smooth-surfaced endoplasmic reticulum (SER). The latter elements are extensive in these cells, whereas ribosomes and elements of the rough-surfaced endoplasmic reticulum are few in number. Some glycogen granules are found in the cytoplasm. Mitochondria (M) are common and are often elongated and have tubular cristae within a dense matrix. The Golgi apparatus, consisting of flattened cisternae and small vesicles, is well-developed. Small vacuoles and irregular cisternae are associated with the Golgi complex.

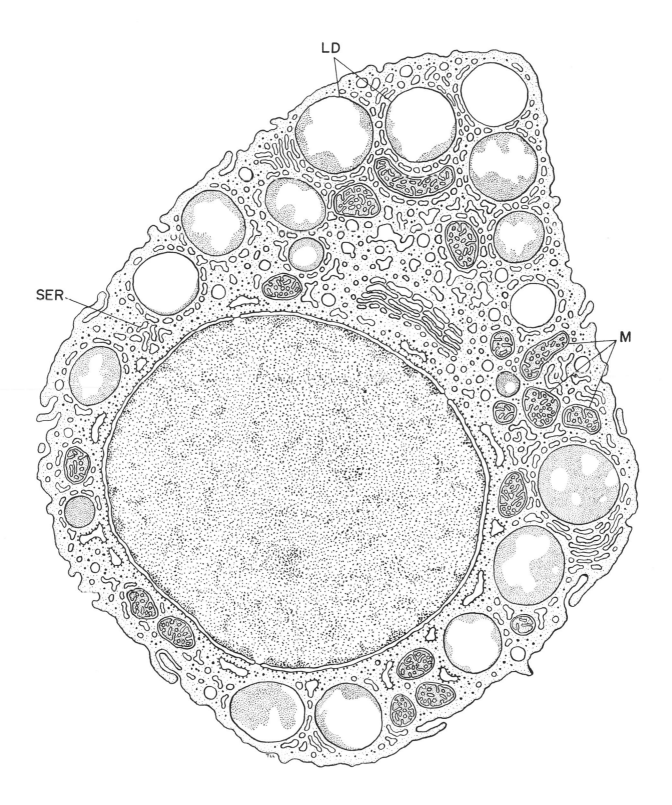

58—MUCOUS MEMBRANE EPITHELIAL CELL

Mucous membranes are composed of incompletely cornified and keratinized cells. Epithelia of this type line the oral cavity, esophagus, vagina, female urethra, and anterior surface of the cornea. Like the skin, the mucosa is a stratified squamous epithelium. The cells, although basically similar in structure to those of the epidermis, differ in a few important respects. These differences are related to the lack of complete cornification (flattening, drying out, and hardening of cells) and keratinization (appearance of the fibrous protein keratin) of the epithelial cells. They contain few keratohyalin granules, and nuclei are retained in the superficial layers. The cells are not as flattened as those of the stratum corneum of the skin.

A cell from the middle zone of the epithelium is illustrated here. It is elongated parallel to the surface of the epithelium and has a number of surface folds and processes. Desmosomes are formed with processes of adjacent cells. The nucleus is centrally located and contains a nucleolus. Most conspicuous in the cytoplasm are bundles of tonofilaments (Tf) running in all directions and into the processes. A Golgi apparatus is found near the nucleus. Mitochondria and some short cisternae of rough-surfaced endoplasmic reticulum are distributed in the cytoplasm. Ribosomes and, especially in the vagina, glycogen granules (Gly) are found. Small, ovoid granules (Gr), the membrane-coating granules, are most common near the cell surface. A rare keratohyalin granule (KG) is located over the tonofibrils.

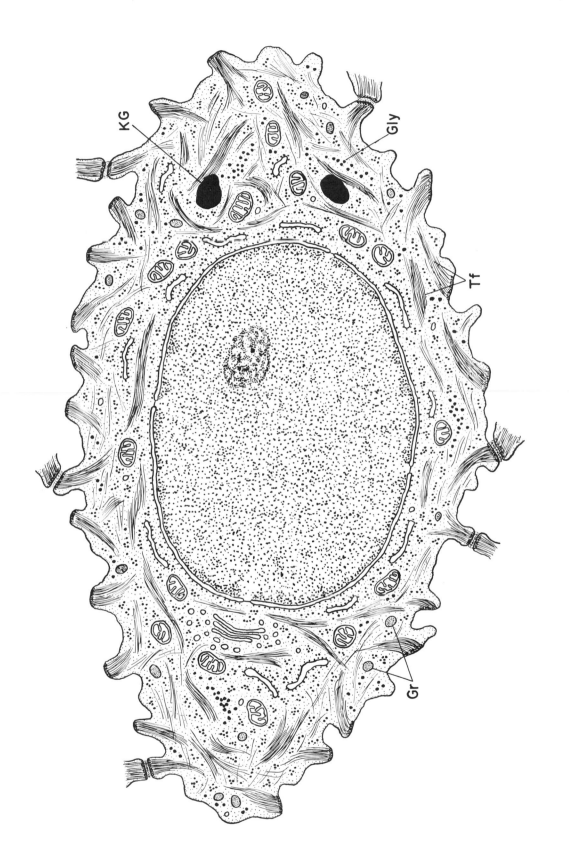

59—MESOTHELIAL CELL

The serous membranes, which include the peritoneum, pleura, and pericardium, are covered by a single layer of mesothelial cells. The mesothelial cells bordering the body cavities are squamous, and the nucleus conforms to the shape of the cell. The apical surface of the cell bears microvilli (Mv), some of which are quite long. Pinocytotic vesicles are found in association with all the cell surfaces including the irregular lateral borders. A small Golgi complex is located near the nucleus. Cisternae of rough-surfaced endoplasmic reticulum, small mitochondria, ribosomes, and a few lysosomes occur in the cytoplasm. There may be relatively large numbers of vesicles (V) in the central regions of the cell. The vesicles perhaps indicate the occurrence of an active exchange of fluids between the mesothelial cell and the serous fluid of the cavities. The microvilli would facilitate this process by increasing the surface area available for exchange of soluble substances. Mesothelial cells may also be capable of taking up particulate materials.

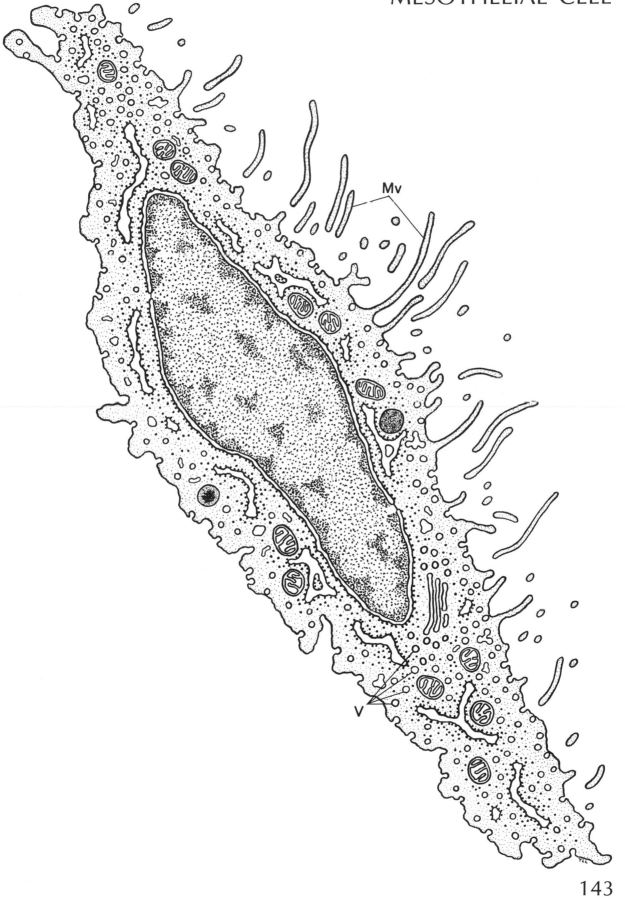

60 — SYNOVIAL CELL

The joint cavities are lined by a loose aggregation of cells that merge gradually with the underlying connective tissue. The lining synovial cells, which appear to be modified fibroblasts, are highly irregular and variable in shape. Villi and cytoplasmic extensions project from the cell surface. The cells have a central round or oval nucleus with a nucleolus and clumps of heterochromatin. A few short, rough-surfaced cisternae and ribosomes occur in the cytoplasm. Stacks of Golgi lamellae are situated near the nucleus. Small vesicles are common at the ends of the lamellae. Large vacuoles containing moderately dense material are found near the Golgi complex and throughout the cell. These organelles may be responsible for the synthesis of the hyaluronate, a mucopolysaccharide, of synovial fluid. Some dense granules resemble lysosomes. Small vesicles, possibly pinocytotic in nature (PV), are seen along the cell surface. Larger vacuoles (Vac) lined with a filamentous layer are in continuity with the plasma membrane and are also free in the cytoplasm. This vesicular surface activity might represent active interchange of other components of the synovial fluid. The synovial cells also contain many mitochondria with transverse cristae. Fine filaments are randomly dispersed in the cell.

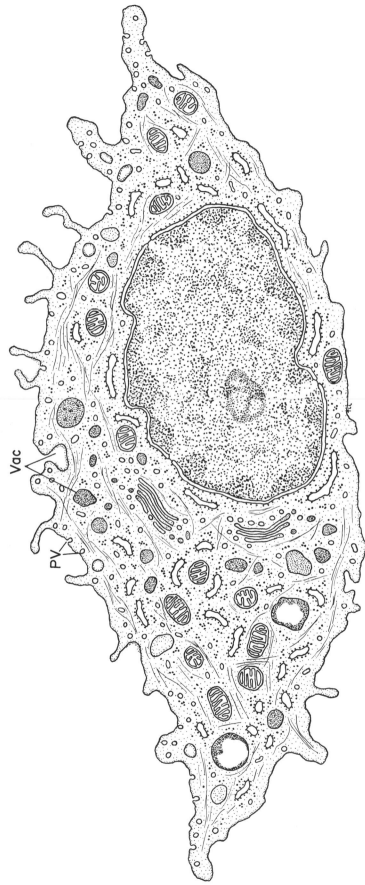

Vac

PV

SKIN AND OTHER EPITHELIA

61—MESENCHYMAL EPITHELIAL CELL

Mesenchymal epithelium is a single layer of squamous cells lining the subdural and subarachnoid spaces, the perilymphatic spaces of the inner ear, and the anterior chamber of the eyeball. These spaces arise as cavities within connective tissue, and the lining mesenchymal epithelium appears to originate through flattening of fibroblast-like cells. Mesenchymal epithelial cells from different regions are basically similar and are uncomplicated in structure. The cells are flattened and interdigitate or overlap adjacent cells. A junctional complex is formed at the apical lateral margins. The cell surfaces are generally smooth with few villi, but pinocytotic invaginations occur on the basal or apical membranes. The nucleus is oval, conforming to the shape of the cell. A small Golgi apparatus is situated near the nucleus. There are only a few short cisternae of rough-surfaced endoplasmic reticulum, but smooth membranes and vesicles are more abundant. Clusters of free ribosomes occur in the hyaloplasm, which has a low density. Mitochondria are round to oval in profile and are uncomplicated in structure. A few lysosomes occur in the cytoplasm.

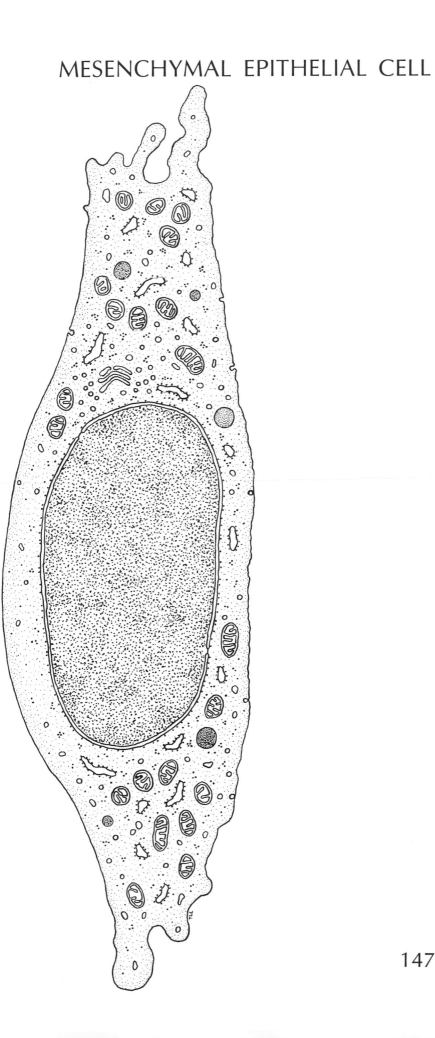

REFERENCES

Barland, P., A. B. Novikoff, and D. Hamerman, 1962. Electron microscopy of the human synovial membrane. J. Cell Biol., 14:207–220.

Cashion, P. D., Z. Skobe, and J. Nalbandian, 1969. Ultrastructural observations on sebaceous glands of the human oral mucosa (Fordyce's "disease"). J. Invest. Derm., 53:208–216.

Cotran, R. S. and M. J. Karnovsky, 1968. Ultrastructural studies on the permeability of the mesothelium to horseradish peroxidase. J. Cell Biol., 37:123–138.

Kiistala, U. and K. K. Mustakallio, 1967. Electronmicroscopic evidence of synthetic activity in Langerhans cells of human epidermis. Z. Zellforsch., 78:427–440.

Munger, B. L., 1965. The cytology of apocrine sweat glands. II. Human. Z. Zellforsch., 68:837–851.

Odland, G. F., 1964. Tonofilaments and keratohyalin. In Montagna, W. and W. C. Lobitz (eds.), The Epidermis. Academic Press, New York, 237–249.

Sagebiel, R. W. and T. H. Reed, 1968. Serial reconstruction of the characteristic granule of the Langerhans cell. J. Cell Biol., 36:595–602.

Terzakis, J. A. 1964. The ultrastructure of monkey eccrine sweat glands. Z. Zellforsch., 64:493–509.

Zelickson, A. S., 1963. Electron Microscopy of Skin and Mucous Membrane. Charles C Thomas, Springfield, Illinois, 173 pp.

DIGESTIVE SYSTEM

DIGESTIVE SYSTEM

62 — TASTE CELL

Two cell types have usually been distinguished in taste buds at the light-microscopic level: the neuroepithelial taste cell and the sustentacular or supportive cell. More recent evidence indicates that the cells of the taste bud represent different functional states, or stages of differentiation, of one basic cell type. This is based on the observations that both sustentacular and neuroepithelial cells are innervated and that there is a constant turnover and replacement of taste cells. The taste cells are narrow and columnar, extending from the basement lamina to the taste chamber. The cell tapers apically to form a prolongation from which several microvilli (Mv) extend. Filaments occupy the core of the microvilli and extend down into the apical cytoplasm. A pair of centrioles (Ce) is sometimes seen in the apex of the cell.

The nucleus occupies the central or basal region of the cell. A Golgi complex overlies the nucleus and may have small dense granules associated with it. Larger dense granules (Gr) are found in the upper region of the cell. These may contribute to the mucopolysaccharide content of the taste chamber. Other cytoplasmic organelles are mitochondria, lysosomes, and short cisternae of both smooth- and rough-surfaced endoplasmic reticulum. Free ribosomes and filaments also occur. Nerve fibers (NF) terminate in close relationship to the taste cells. The fiber may be embedded in or even enveloped by the taste cell. The nerve ending contains some small mitochondria and small vesicles.

The taste cell is most likely stimulated by adsorption of substances to the membrane of the microvilli, producing an electrical response. The impulse is carried from the receptor pole of the cell to the basal region and transmitted to the nerve possibly by an acetylcholine mechanism. There are four basic qualities of taste: sweet, salt, bitter, and acid. The variety of tastes perceived results from combinations of the basic tastes, and olfactory and somatosensory contributions.

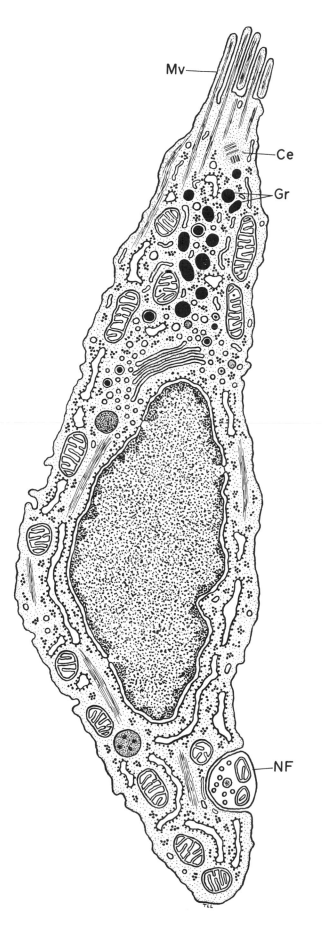

63 — TOOTH: AMELOBLAST

Secreting ameloblasts (ganoblasts) of the enamel organ are tall, extremely narrow cells of ectodermal origin first situated in a layer next to the odontoblasts but then separated from them by the enamel and dentin. An oval nucleus with a nucleolus occupies the lower third of the cell. Some short, basal microvilli (Mv) extend into the stratum intermedium at the base of the cell. A cytoplasmic web of fine filaments (Fl) runs immediately below the plasma membrane. A cluster of mitochondria and masses of glycogen granules (Gly) are situated in the basal cytoplasm between the cytoplasmic web and nucleus.

A large Golgi apparatus (G) is located in the supranuclear portion of the cell. The complex is oval in shape and is composed of peripheral stacks of membranous cisternae surrounding a central core of cytoplasm. The Golgi cisternae are flattened in the inner regions of the stacks but are more dilated and sacculated on the outer sides. The cytoplasm enclosed by the Golgi membranes contains small, membrane-bounded secretory granules, vesicles, and both smooth- and rough-surfaced cisternae of endoplasmic reticulum.

Dense granules like those in the Golgi region are common in the distal portion of the cell. The content of the granules is similar to the deposits of dense material in the extracellular spaces adjacent to the distal end of the cell. Enamel crystallites first appear in contact with the dense extracellular material. It is thought, therefore, that the ameloblasts elaborate the enamel matrix, which subsequently calcifies. Fully developed enamel consists mostly of apatite crystals and little organic material, and it is the hardest substance occurring in the body. Small vesicles are also common in the apex of the cell, and some are in continuity with the plasma membrane. Larger secretory granules and multivesicular bodies, some with a dense matrix, are found in the vicinity of the Golgi complex.

The cytoplasm lateral to the Golgi apparatus is largely filled with elongated cisternae of rough-surfaced endoplasmic reticulum that branch and anastomose. Free ribosomes occurring in clusters are common. Microtubules (Mt) and bundles of fine filaments (Fl) run parallel to the long axis of the cell. These filaments extend into the basal web of filaments and into a similar web at the distal pole of the cell.

An ameloblastic process called Tomes' process (top of Plate) extends from the tip of the cell beyond the terminal web region into the enamel. The process has no major organelles but contains microtubules, secretory granules (SG), and vesicles. When the enamel is formed, the ameloblasts atrophy and disappear.

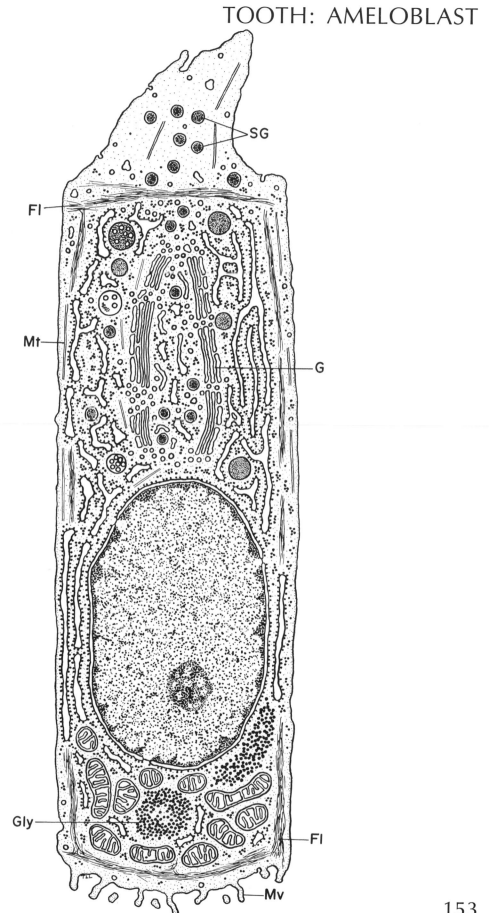

SG

Fl

Mt

G

Gly

Fl

Mv

64—TOOTH: ODONTOBLAST

The odontoblasts are of mesodermal origin and cover the dental papilla which later becomes the pulp cavity and underlies the dentin. They are believed to be responsible for the formation of the organic matrix of dentin called the predentin. Predentin consists of collagen fibers in a ground substance containing a mucopolysaccharide. The predentin then becomes calcified to form dentin. The cells are highly columnar with a basally situated nucleus. The rough-surfaced endoplasmic reticulum is extensive and occupies much of the cytoplasm. Many of the cisternae are elongated, while others are shorter and more dilated. The central region of the cell is occupied by an extensive Golgi apparatus (G). The Golgi complex is composed of several stacks of membranous cisternae and small vesicles; dense granules (Gr) are situated near it. There are granules and mitochondria throughout the cell. Cytoplasmic filaments are common and, at the apex of the cell, form a terminal web.

An odontoblastic process (top of Plate) extends from the terminal web area into the dentin. These processes, or Tomes' fibers, fill the dentinal tubules, which are minute canals radiating from the pulp cavity to the periphery of the dentin. The cytoplasmic process is devoid of most organelles but contains a few dense granules, filaments, and microtubules. Vesicles are abundant, and many are fused with the plasma membrane.

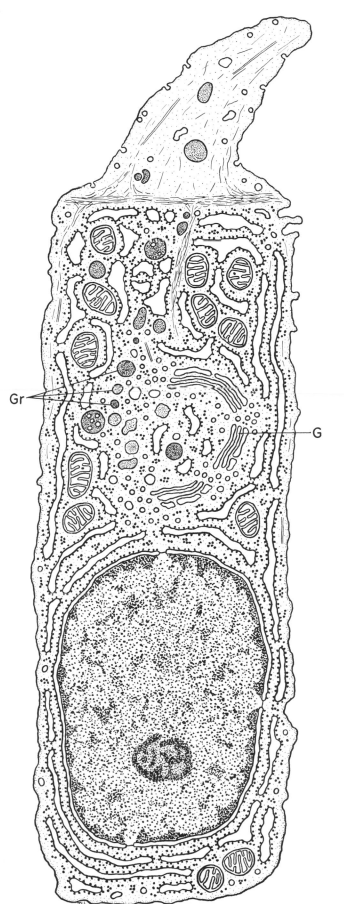

Gr

G

65 — SALIVARY GLAND: MUCOUS CELL

A number of glands open into the oral cavity and secrete saliva. These include many small glands in the mucosa or submucosa, and three pairs of large salivary glands: the parotid glands, the mandibular or submaxillary glands, and the sublingual glands. The glands contain different proportions of mucous and serous cells. The smaller glands may be purely mucous, or purely serous (glands of von Ebner), or mixed. The parotid glands are composed of serous cells; the submandibular glands are mostly serous with some mucous cells; and the sublingual glands are mixed, with a preponderance of mucous cells.

Mucous cells in the glands of the oral cavity are similar to mucus-producing cells found elsewhere (goblet cell, Plate 80; mucous cell of respiratory epithelium, Plate 92). The cytoplasm is crowded with electron-lucent mucous granules (MG). The nucleus is confined to the base of the cell. The Golgi apparatus is found near the nucleus, whereas cisternae of rough-surfaced endoplasmic reticulum occupy the cytoplasm at the cell base and between the mucous droplets. Free ribosomes are abundant, but only a few mitochondria and lysosomes occur in the basal cytoplasm. Some short microvilli extend from the cell surface into the gland lumen. Mucous cells elaborate a viscid secretion consisting predominantly of mucin.

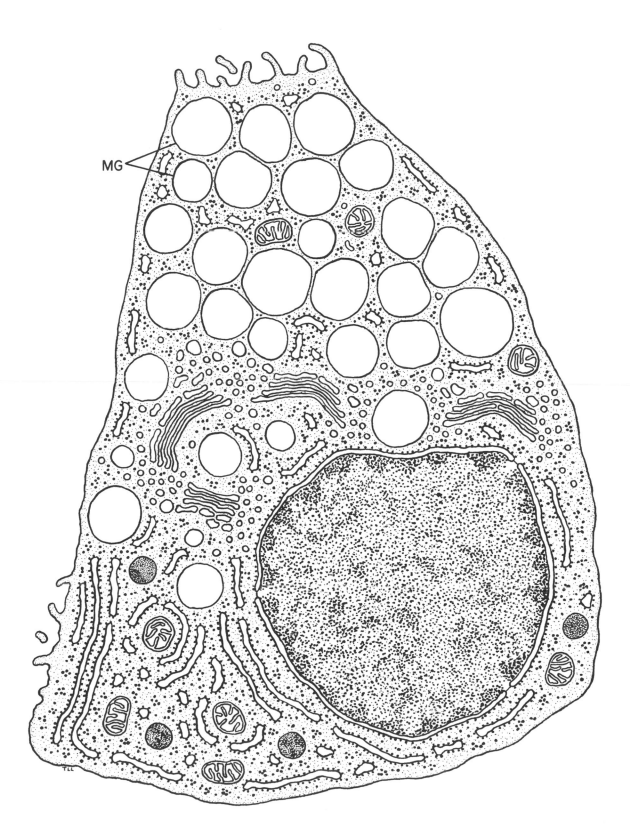

DIGESTIVE SYSTEM

66—SALIVARY GLAND: SEROUS CELL

The serous cells are most commonly located at the terminal ends of the secretory portions of the glands. In the sublingual glands especially, they are flattened and surround the mucous cells to form serous demilunes. The secretions of these cells are carried to the lumen through secretory capillaries between the mucous cells. The serous cells lining the acini are pyramidal with a basally located nucleus containing a nucleolus. Microvilli extend from the apical surface into the lumen. The cytoplasmic structure is similar to that of most other cells producing a protein for export. Elongated cisternae of rough-surfaced endoplasmic reticulum fill the basal region of the cell. The Golgi apparatus, above the nucleus, is extensive, as in most actively secreting cells. Vacuoles containing material of low to medium density (condensing vacuoles, CV) occur in the Golgi region. The apex of the cell is filled with dense secretory or zymogen granules (ZG) that contain the enzyme ptyalin. During secretion, the granules fuse with the apical plasma membrane and with each other. Mitochondria are distributed throughout the cell, and free ribosomes occupy the hyaloplasm. (Compare the serous cell with the pancreatic acinar cell on Plate 84.)

The serous glands elaborate a watery liquid containing ptyalin and salts. Saliva is a mixture of the secretion of all the glands and includes water, mucin, salts, protein, and ptyalin. It serves to lubricate the oral cavity, to provide moisture, and to break down starch to soluble carbohydrates through the action of ptyalin.

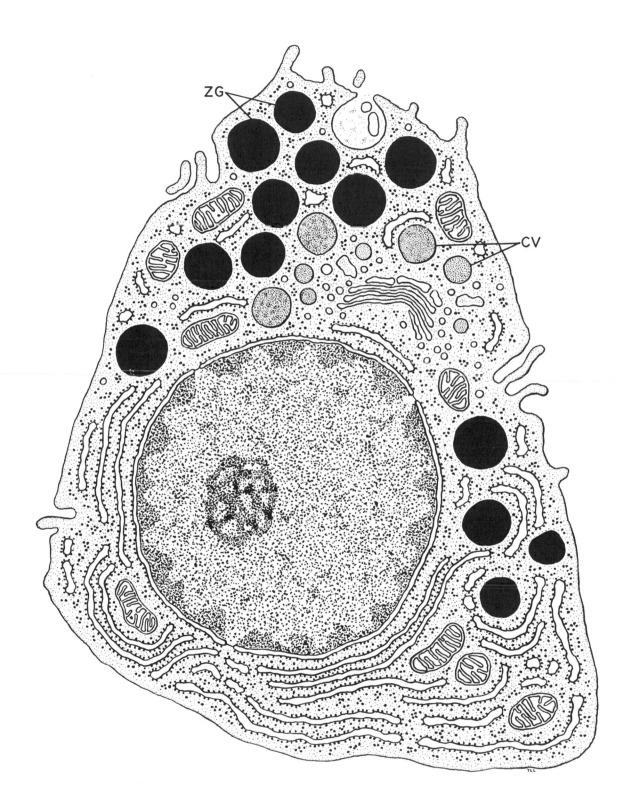

DIGESTIVE SYSTEM

67 — SALIVARY GLAND: INTERCALATED DUCT CELL

The intercalated duct arises from the acinus of the salivary gland and is composed of cuboidal cells enclosing a small lumen. The cells are joined laterally by junctional complexes. A few short microvilli occur on the luminal surface, there are some lateral folds, and the basal surface is relatively flat. The round nucleus occupies a major portion of the cell. Cytoplasmic structure is uncomplicated. A Golgi apparatus occurs near the nucleus, but elements of endoplasmic reticulum are sparse. Some mitochondria and an occasional lysosome or multivesicular body are found in the cytoplasm. Free ribosomes and bundles of filaments occur as well. Some cells, especially those nearest the acinus, may contain a few small secretory granules.

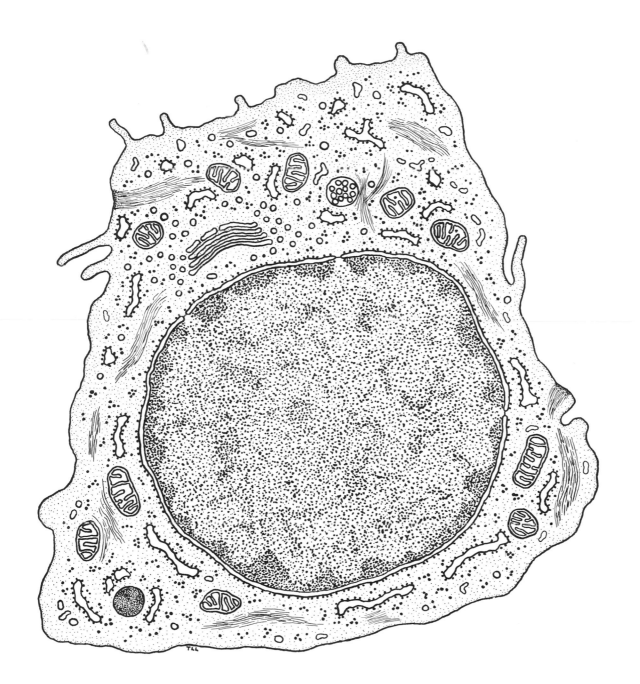

68 — SALIVARY GLAND: STRIATED DUCT CELL

The striated ducts of the salivary glands are continuous with the intercalated ducts. The cells are columnar with a round central nucleus. The ducts derive their name from the prominent infoldings of the basal plasma membrane. The cytoplasm between the folds is filled with elongated mitochondria (M) oriented in the long axis of the cell. The cristae of the mitochondria are tightly packed, and the matrix is dense. The basal infoldings of duct cells are similar to those of distal convoluted tubule cells of the kidney (Plate 101), and it has been suggested that the duct cells also may function in transport of water and ions.

The apical surface of the cell has some stubby microvilli protruding into the lumen. Folds and interdigitations occur laterally. The apical cytoplasm is crowded with short, smooth-surfaced tubules and with vesicles (V) that have clear contents. The perinuclear cytoplasm contains a small Golgi apparatus, some spherical mitochondria, and a few lysosomes. Rough-surfaced cisternae of endoplasmic reticulum are not abundant, but some free ribosomes occur. Bundles of filaments are common. Secretory granules have been observed in the apical cytoplasm of striated duct cells in the submandibular gland.

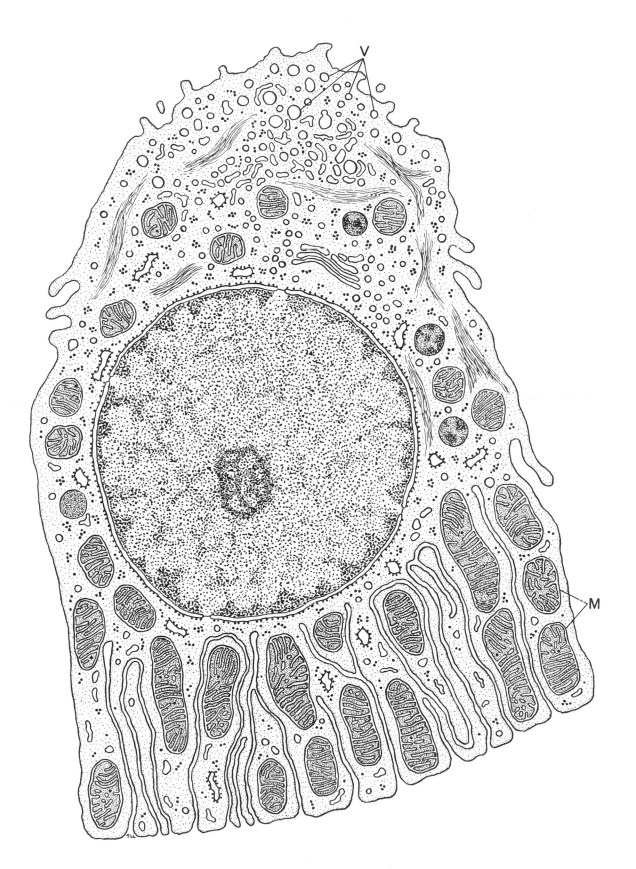

DIGESTIVE SYSTEM

69—STOMACH: PARIETAL CELL

Parietal or oxyntic cells are pyramidal or triangular and occur singly in gastric glands. One of the striking morphological specializations of this cell is secretory canaliculi (Ca) that extend from the apex of the cell and pass lateral to the nucleus almost to the base of the cell. The canaliculus takes the form of a sinuous tubular channel, or it may form a deep indentation that is completely concentric with the nucleus. Numerous microvilli extend into the lumen of the canaliculus. In some places, strands of cytoplasm extend across the lumen. The canaliculi open into a common outlet that is continuous with the lumen (Lu) of the gland. Thus, the extensive surface area of the canaliculi is exposed to the glandular lumen. Microvilli also occur on the surface of the common opening and apical region of the cell. The lateral surfaces are relatively smooth in contour, but the basal surface is thrown into villi or plications.

Another unusual structural feature of this cell is a system of cytoplasmic tubules (Tu). These tubules are sometimes so extensive that they crowd the other organelles into basal or lateral positions. The tubules are limited by smooth membranes, and the enclosed space is electron-lucent. In places these tubules may be continuous with the surface plasma membrane of the canaliculus, in which case they should be regarded as complex invaginations of plasma membrane instead of smooth-surfaced endoplasmic reticulum. Mitochondria are relatively large, and spherical or oval in shape. Cristae are closely packed and traverse over half the width of the organelle. Lysosomes (Ly) are present, as well as a few short cisternae of rough-surfaced endoplasmic reticulum. Free ribosomes occur in the cytoplasm, and a small Golgi apparatus is located near the base of the cell, sometimes in an infranuclear position. Parietal cells secrete the hydrochloric acid of gastric juice, but the relationship of the structural specializations to acid secretion is not well understood. One possible mechanism is that sodium chloride is secreted into the canaliculi, whereupon the sodium is exchanged for hydrogen ions across the membrane.

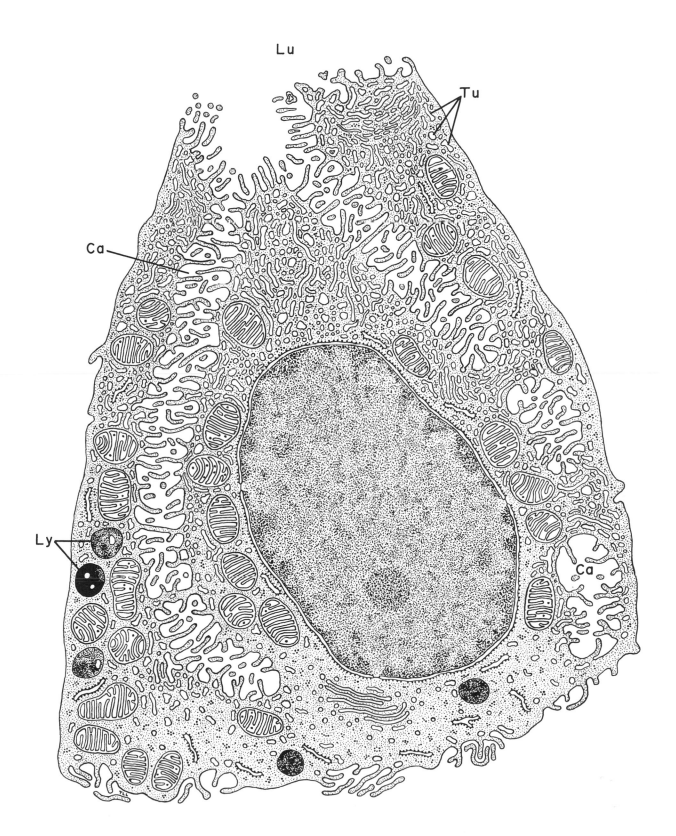

DIGESTIVE SYSTEM

70 — STOMACH: SURFACE MUCOUS CELL

The lumen of the stomach and the gastric pits are lined by the surface mucous cells. These cells are columnar or cuboidal, with the nucleus in the basal region of the cell. A few short microvilli project from the apical surface, which is covered with fine filamentous material. Apically, the lateral plasma membranes are smooth in contour, but they are markedly interdigitated basally. The mucous granules (MG) are concentrated in the upper portion of the cell and are spherical or ovoid. Most are extremely dense, but in some cases only the central core of the granule is dense. Some of the granules are situated immediately beneath the plasma membrane between microvilli, or the membrane of the granule is continuous with the plasma membrane and forms an opening. Mitochondria are most abundant in a zone beneath the mucous granules. One or more elaborate Golgi complexes occur above or at the sides of the nucleus. This organelle is composed of a stack of membranous lamellae and many small vesicles. Vacuoles in the vicinity of the Golgi apparatus contain moderately dense material. Some larger granules may occur in this region, but they are not as dense as the apical granules. The granules of medium density may represent intermediate stages in the formation of mucous granules. Free ribosomes occur in the cytoplasm, but cisternae of rough-surfaced endoplasmic reticulum are few in number.

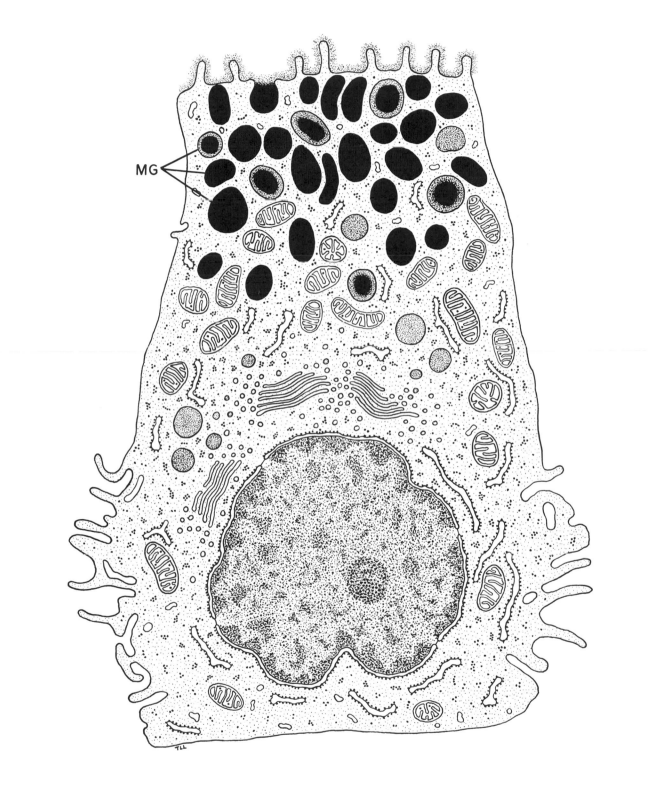

MG

71—STOMACH: MUCOUS NECK CELL

Mucous neck cells occur in the neck of the gastric glands of the fundus and extend into the base of the glands in the pyloric and cardiac regions. These cells are usually columnar but may be club- or flask-shaped. The mucous secretory granules (MG) of the neck cell differ from those of the surface mucous cell (Plate 70). In the mucous neck cell, the granules are larger, more uniformly spherical, and of lower density. Instead of being confined to the apex of the cell, they have a wider distribution in the cytoplasm. The nucleus, containing a nucleolus, is located in the basal region of the cell. The supranuclear Golgi apparatus consists of a few lamellae, vesicles, and some larger vacuoles containing material similar in appearance to that of mature mucous granules. A moderate number of elongated cisternae of rough-surfaced endoplasmic reticulum is present. The cytoplasm also contains free ribosomes and mitochondria. A few short microvilli project into the lumen, and some interdigitations occur on the lateral surface. The mucous cells of the gastric mucosa secrete mucus, which forms a layer over the epithelium. This layer may be protective, preventing autodigestion by the gastric juice.

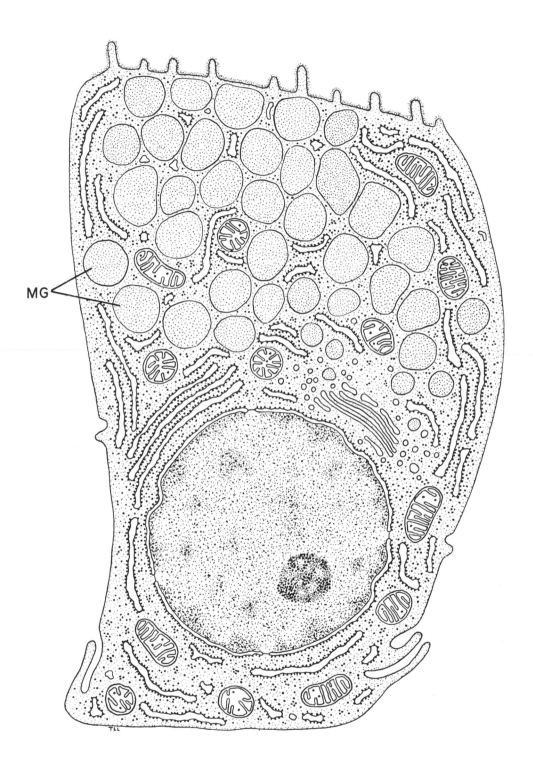

MG

72—STOMACH: CHIEF CELL

Chief or zymogenic cells occupy the greater part of the bases of gastric glands in the fundus. The cells are cuboidal or low-columnar in shape and have a basally situated nucleus. The nucleus contains a prominent nucleolus. Finely filamentous material covers the apical plasma membrane. A few microvilli project into the glandular lumen. The supranuclear portion of the cell contains numerous zymogen granules (ZG) that are thought to contain pepsinogen. Most of the spherical granules are of very low density, but a few are moderately dense or have a dense core surrounded by a lucent rim. The granules are limited by a membrane, and at the surface of the cell the granule membrane may be fused with the plasma membrane, thereby allowing the contents to escape into the lumen. A Golgi complex occurs above the nucleus and has some small vacuoles associated with it. The content of the Golgi vacuoles is similar in appearance to that of the mature zymogen granules. Mitochondria and endoplasmic reticulum are concentrated in the basal region of the cell. The endoplasmic reticulum is extensive and composed of elongated, tubular and lamellar cisternae. Ribosomes occur on the membranes of the endoplasmic reticulum as well as free in the cytoplasm.

Gastric juice is composed of water, hydrochloric acid, salts, mucin, and the enzymes pepsin and renin. Chief cells appear to be protein-secreting cells and would thus produce pepsin and renin. They may also elaborate gastric intrinsic factor, which facilitates the absorption of vitamin B_{12}.

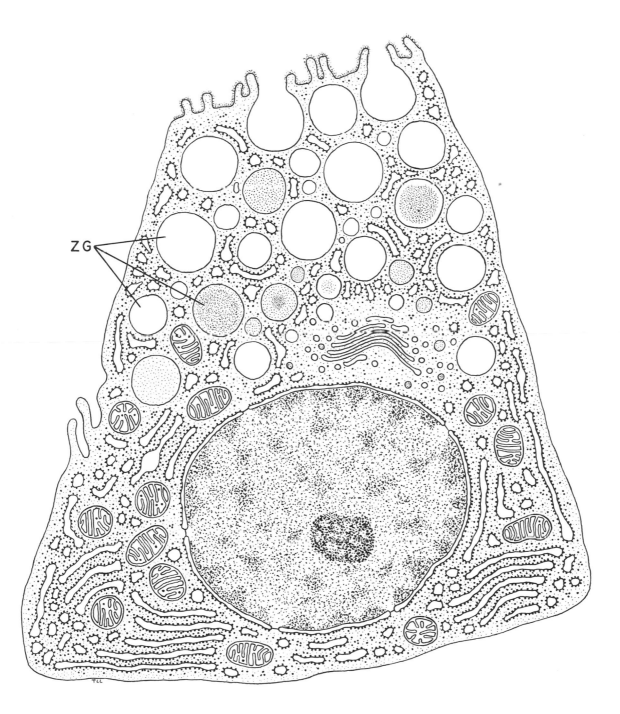

ZG

73—STOMACH: ARGENTAFFIN CELL

Argentaffin or enterochromaffin cells occur singly at the base of the cardiac, fundic, and pyloric glands, in the intestinal crypts, and in the appendix and colon. It is the most commonly occurring endocrine cell in the gastrointestinal tract (see Plate 74). The cells are polygonal or flattened with the broad end facing the basement lamina. Usually they do not extend to the surface of the epithelium, although a few cells may send a narrow prolongation bearing microvilli to the lumen. Argentaffin cells have a central nucleus that is irregular or indented in contour. Spherical, dense granules are abundant in the cytoplasm, especially basally. The granules are bounded by a loose-fitting membrane. A Golgi apparatus (G) occurs near the nucleus, often in the pole of the cell facing the basement lamina. Associated with the Golgi apparatus are vacuoles containing small granules. The granule content varies in density from moderate to opaque and may be a precursor of the mature argentaffin granule. Some spherical or oval mitochondria, short cisternae of rough-surfaced endoplasmic reticulum, and free ribosomes are present. The argentaffin cell granules contain 5-hydroxytryptamine (serotonin), which stimulates smooth muscle contraction.

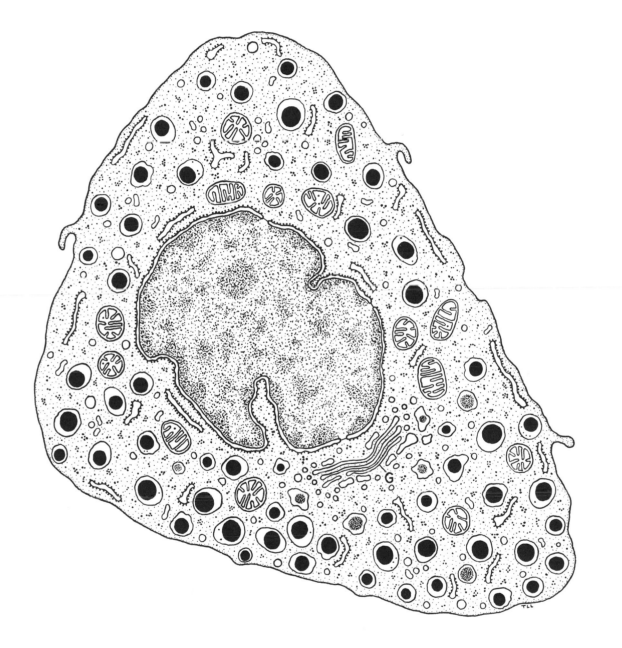

DIGESTIVE SYSTEM

74—STOMACH: GASTROINTESTINAL ENDOCRINE CELL

Several types of endocrine cells have been identified in the epithelium of the gastrointestinal tract. These types, which are identified on the basis of structural and functional differences, are (1) argentaffin cells, which produce serotonin or 5-hydroxytryptamine (Plate 73); (2) chromaffin cells, which are thought to produce catecholamines (Plates 149, 150); (3) enteroglucagon-producing or intestinal A cells, comparable to the alpha cells of the pancreas (Plate 151); (4) intestinal D cells, resembling the delta cells of the pancreas (Plate 153); and (5) gastrin-producing cells.

The presumed gastrin-producing cell has been found in the pyloric region of the stomach and, less commonly, in the cardia and upper part of the duodenum. The cells are oval and generally extend the full height of the epithelium. The narrow free surface bordering the gland lumen is provided with a few microvilli (Mv). The nucleus is found at the base of the cell and has a peripheral rim of condensed chromatin. Much of the cytoplasm is occupied by membrane-bounded secretory granules (SG). The granules have a fairly uniform diameter of 300 to 500 mμ, but the density of their content differs greatly. Some of the contents vary from lucent to finely flocculent, whereas others are moderately dense or opaque. Ribosomes are abundant in the cytoplasm between organelles. There are a number of flattened cisternae of rough-surfaced endoplasmic reticulum, especially around the nucleus. A Golgi apparatus is positioned near the nucleus. The cytoplasm also contains mitochondria and some small bundles of filaments. Evidence does not yet seem conclusive that this cell produces gastrin, which stimulates the flow of gastric juice, but the cell is affected by fasting and feeding (Forssmann et al.).

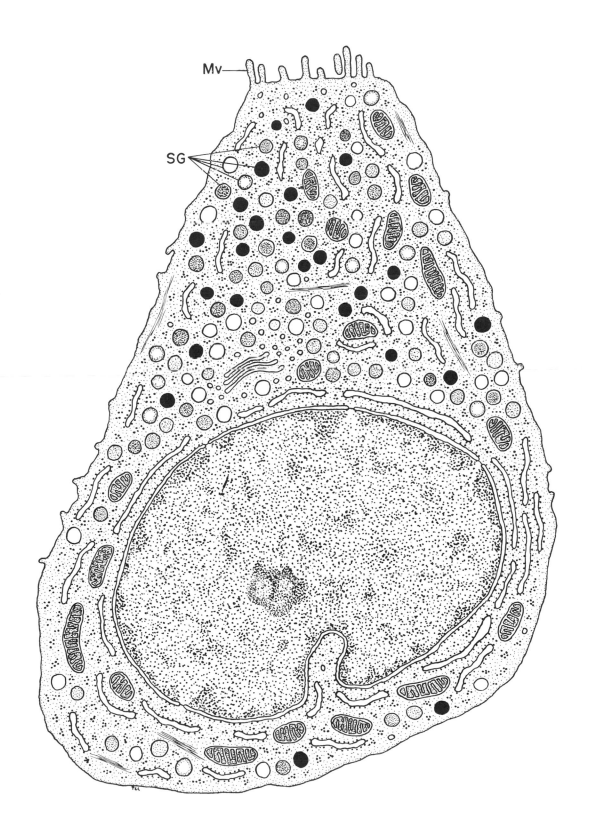

75—INTESTINES: BRUNNER'S GLAND CELL

These cells comprise the Brunner's glands found in the submucosa of the duodenum. The cells are cuboidal to low-columnar in shape; the basal nucleus contains a nucleolus and evenly dispersed chromatin material. Microvilli extend from the free surface of the cell. Laterally and apically between adjacent cells, small intercellular channels or secretory canals (Ca) may occur. Toward the base, complex interdigitations are found. The gland cells contain a well-developed endoplasmic reticulum that is most extensive in the basal portion of the cell. The rough-surfaced cisternae range from short tubular profiles to elongated cisterns that extend in parallel. Free ribosomes occur singly and in clusters, in the cytoplasm. These cells have an extensive Golgi system (G) that is dispersed over a wide area above the nucleus. Flattened, smooth membranous lamellae occur in parallel aggregations. The outermost cisternae are flat and empty, while the innermost are distended by material similar to that comprising early secretory granules. Vesicles are dispersed around the margins of the lamellar stacks, especially laterally and distally. Vesicles are also abundant near rough-surfaced cisternae of endoplasmic reticulum that face the Golgi region, and some appear to bud off from the cisternae. Vacuoles (Vac) containing secretory material are numerous along the inner face of the Golgi. Small spherical secretory granules (SG) with a dense content are found in the Golgi zone. The granules are larger in the distal areas of the cell and accumulate in the apex. Mitochondria are large, and round or oval in shape. Some are partially surrounded by cisternae of endoplasmic reticulum. Multivesicular bodies are sometimes found in the Golgi zone. Brunner's glands secrete a clear, viscous, alkaline fluid that may protect the duodenal mucosa from the action of acid gastric fluid. It has been suggested that the protein component of the secretory product is synthesized in the rough-surfaced endoplasmic reticulum, while synthesis of a carbohydrate component and its combination with the protein moiety may occur in the Golgi complex.

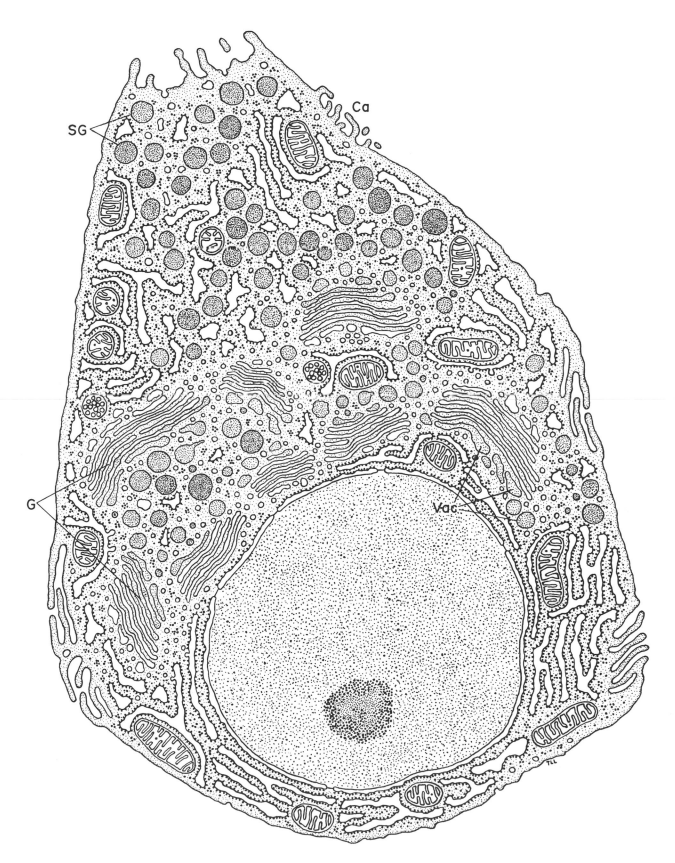

DIGESTIVE SYSTEM

76—INTESTINES: INTESTINAL EPITHELIAL CELL

The intestinal epithelial cell is columnar and lines most of the luminal surface of the small intestine. Microvilli (Mv) extend from the free surface and form the striated border of the cell. The surfaces of the microvilli are covered by a coating of filamentous material believed to be a mucopolysaccharide. Complex interdigitations and a few vesicular invaginations occur on the lateral surfaces.

The nucleus, containing a prominent nucleolus, is situated in the lower portion of the cell. The basal cytoplasm contains mitochondria, free ribosomes, and lipid droplets (LD). The Golgi apparatus (G), consisting of membranous lamellae, small vesicles, and some vacuoles, occurs above the nucleus. Mitochondria and endoplasmic reticulum are found in the supranuclear region. The mitochondria are large, oval to elongated in shape, with a moderate number of cristae extending transversely across the organelle. The rough-surfaced endoplasmic reticulum (ER) consists of short and long membranous cisternae bearing ribosomes on their outer surfaces. The elongated cisternae usually run parallel to the long axis of the cell, and some closely follow the mitochondrial contours. The smooth endoplasmic reticulum (SER) consists of round or short tubular profiles that may branch and anastomose. These are most common immediately below the terminal web area. Some of the smooth tubules are in continuity with rough-surfaced cisternae. Other structures in the supranuclear cytoplasm are microtubules, free ribosomes, and lysosomes (Ly). Lysosomes have dense contents or contain many small vesicles.

The terminal web area (TW) is a narrow zone beneath the microvillous border from which most of the general cytoplasmic organelles are excluded. Fine filaments are abundant in this region and extend parallel to the surface. A bundle of filaments occupies the core of each microvillus and extends basally to intermingle with the terminal web filaments. Pinocytotic invaginations or apical pits occur at the bases of some of the microvilli. In the terminal web area are spherical or tubular membrane-bounded bodies or apical vesicles (AV), which may be derived from the apical pits (AP). A few lysosomes may be found in the terminal web zone, and tubules of smooth endoplasmic reticulum extend into this region.

The intestinal epithelial cells are responsible for the absorption of amino acids, monoglycerides and fatty acids, and monosaccharides, the end products of the digestive process. Because lipid can be seen with the electron microscope, more is known about the morphology of lipid absorption (Plate 77) than about that of the other end products. The microvilli of the brush border greatly amplify the absorptive surface. The brush border contains adenosine triphosphatase, which may be associated with the active transport of substances into the cell. The border, in addition, contains other enzymes, including invertase, maltase, and various peptidases, and is thus also responsible for the terminal hydrolysis of carbohydrates and proteins.

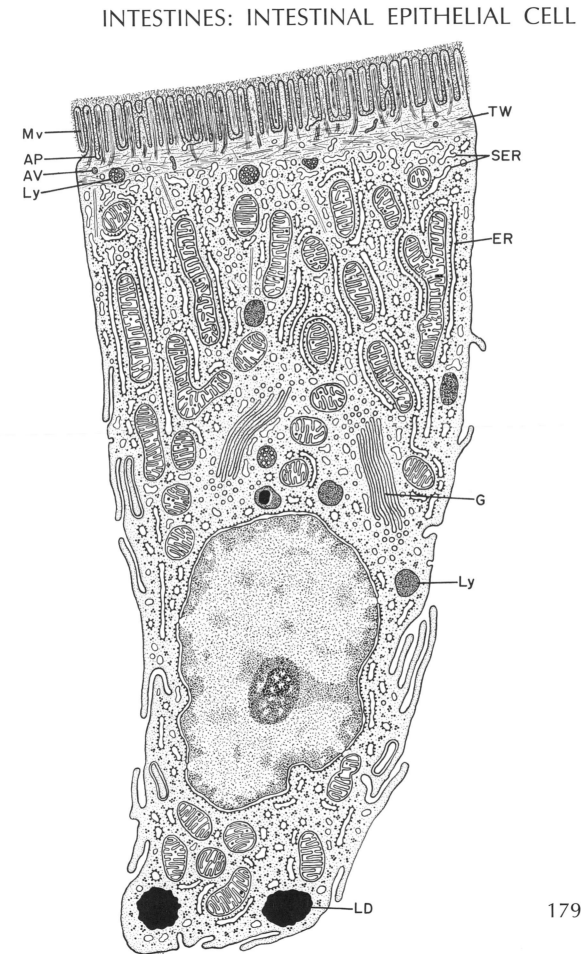

Mv
AP
AV
Ly

TW
SER
ER
G
Ly
LD

179

DIGESTIVE SYSTEM

77—INTESTINES: INTESTINAL EPITHELIAL CELL—FAT ABSORPTION

Following fat administration to an animal, no morphological changes can be detected in the microvilli of the intestinal epithelial cells. Similarly, apical pits and apical vesicles are unchanged in number or structure except for the rare occurrence of a fat droplet in an apical pit. Striking alterations, on the other hand, occur in the endoplasmic reticulum. The smooth endoplasmic reticulum (SER) becomes much more prominent and occupies a greater proportion of the apical cytoplasm, while cisternae of rough-surfaced endoplasmic reticulum become shorter and reduced in number. Fenestrations occur along the margins of the rough cisternae. Lipid droplets are found in tubules, bulbous expansions, and isolated vesicles of the smooth endoplasmic reticulum. Continuities persist between the smooth and rough cisternae, and sometimes a lipid droplet is seen in a rough cisterna.

The Golgi apparatus (G) is also conspicuously enlarged during fat absorption. Initially, Golgi vacuoles contain several fat droplets the size of those within the smooth endoplasmic reticulum. The small droplets apparently fuse to produce larger droplets. Eventually, each Golgi vacuole contains one large droplet of liquid. Vesicles containing single lipid droplets are abundant along the lateral margins of the cell. The intercellular spaces are enlarged and contain lipid droplets or chylomicra (Chy). A few coated pits border the intercellular spaces and may represent vesicles that have fused with the plasma membrane and discharged their contents.

Correlation of biochemical findings with the morphological observations has led to the following theory of intestinal triglyceride absorption. Large droplets of emulsified triglycerides in the intestinal lumen are attacked by pancreatic lipase, which releases free fatty acids and monoglycerides. These products of hydrolysis combine with conjugated bile salts to form a micellar solution. Monoglycerides and fatty acids then diffuse from the micelle through the plasma membrane of the microvilli into the terminal web area. They are then incorporated by the smooth endoplasmic reticulum, synthesized into triglycerides, and segregated as fat droplets within the lumina. The vesicles of smooth endoplasmic reticulum may transport the enclosed lipid to the lateral cell membrane where the vesicles discharge the lipid as chylomicra to the intercellular space. The intercellular chylomicra are then available for absorption by the lacteals.

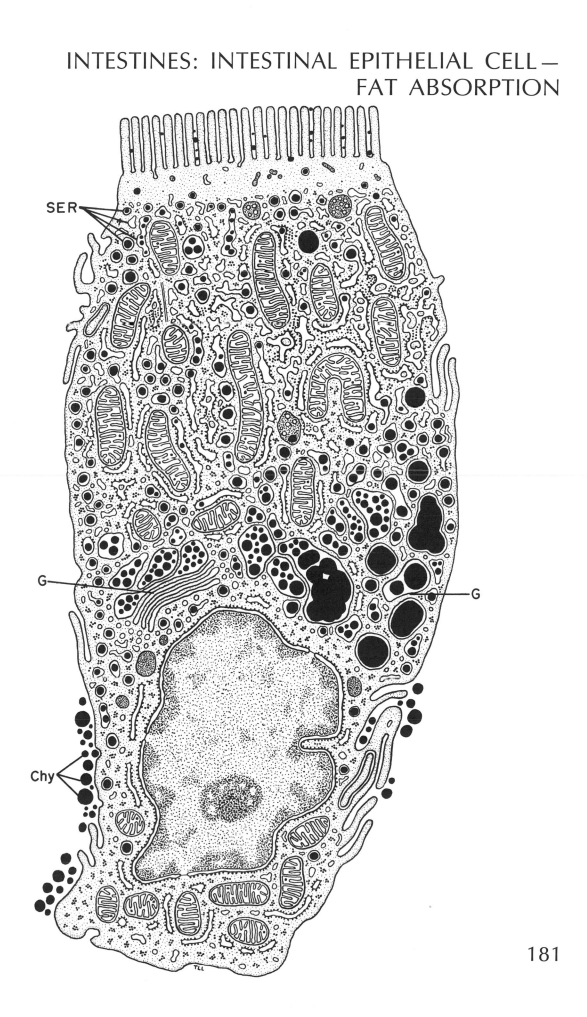

SER

G

G

Chy

181

DIGESTIVE SYSTEM

78—INTESTINES: INTESTINAL EPITHELIAL CELL—JUNCTIONAL COMPLEX

A characteristic junctional complex occurs between epithelial cells forming glands and lining cavitary organs. This complex has been found in the mucosal epithelia of the stomach, intestine, gallbladder, uterus, and oviduct and in the glandular epithelia of the liver, pancreas, parotid, stomach, and thyroid. It is encountered in the epithelia of pancreatic, hepatic, and salivary ducts and between epithelial cells of the nephron including proximal and distal convoluted tubules and collecting ducts. The junction is composed of three elements: the *zonula occludens* (tight junction), *zonula adherens* (intermediate junction), and *macula adherens* (desmosome). Although the complex differs somewhat in various organs, it is probably of general occurrence in epithelia.

The junctional complex of the intestinal epithelium will be described in detail here but is not included in all illustrations of other epithelia. The unit membrane structure of the plasma membrane (PM) is shown, in which two dense leaflets are separated by a clear space. The cell surface bears an extraneous coat (EC) rich in polysaccharides. This layer is composed of fine filamentous material that appears to be attached to the outer leaflet of the unit membrane. The first element of the complex is the *zonula occludens* (ZO), which begins where the surface plasma membrane is reflected inward. The adjacent plasma membranes fuse to obliterate the intercellular space. An intermediate line is formed by fusion of the outer leaflets of the plasma membranes. There may be a slight increase in the density of the cytoplasm adjacent to the junction, but there are no associated fibrils. The tight junction is 0.2 to 0.5 μ in depth and forms a continuous belt around the apical perimeter of the cell.

At the base of the tight junction, the membranes diverge and the second element of the complex, the *zonula adherens* (ZA), begins. The membranes extend in parallel, separated by an intercellular space of 200 Å. Dense material occurs in the subjacent cytoplasmic matrix. The dense material has a fibrillar texture and is continuous with the terminal web (TW), into which the filamentous rootlets (Rt) of the microvilli (Mv) extend. A plate-like density often occurs in the fibrillar material running parallel to the plasma membrane and separated from it by a lighter space. The intermediate junction is 0.3 to 0.5 μ deep and also forms a complete belt around the cell.

The third element of the junction is the *macula adherens* (MA) or desmosome. Between the intermediate junction and the desmosomes is a short region with an irregular intercellular space. Vesicles (V) and invaginations of the plasma membrane occur in this area but not in association with the membranes forming the junctional elements. At the level of the desmosome, the intercellular space is 240 Å across and is occupied by a disk of moderately dense material bisected by a denser intermediate line. Plaques of dense material are located parallel to the inner leaflet of the plasma membrane and separated from the membrane by a space of lower density. Bundles of cytoplasmic fibrils converge on the inner aspect of the desmosomal plaques. These fibrils also occur in the cytoplasm beneath the terminal web and are coarser and more distinct than those of the terminal web. The desmosome is 0.2 to 0.3 μ long, and is a discontinuous, button-like structure.

Regarding the functions of the junctional complex, all three elements appear to be involved in cell-to-cell attachment. The *zonula occludens* is the last element of the complex to break down under tension. The *zonula occludens* is also a diffusion barrier, being impermeable to macromolecules and probably also to water, ions, and small water-soluble molecules. Finally, cells joined by tight junctions are electrically coupled, indicating that this junction serves as a low resistance pathway for the spread of ionic current between cells. Besides serving as a pathway for ions, there is evidence that larger molecules can pass from cell to cell through the *zonula occludens*.

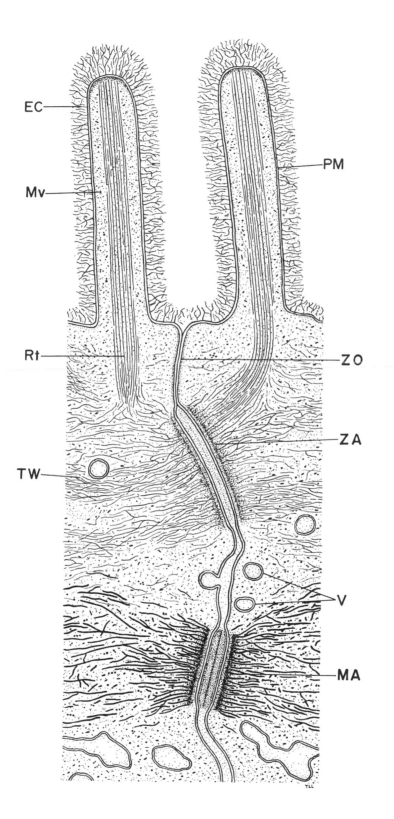

79 — INTESTINES: PANETH CELL

Paneth cells are pyramidal cells found at the bases of the crypts of Lieberkühn of the small intestine. The supranuclear portion of the cell is filled with large spherical secretory granules (SG). There are variations in the texture and density of the granules, depending partly on the fixative employed. Single cells usually contain predominantly one type of granule, although several of the variations are shown here. In the mouse, the granules are heterogeneous in internal structure, with a halo surrounding a central core. The halo may be less dense or more dense than the core, and it is not separated from the core by a membrane. At the apex of the cell, the membrane limiting the granule fuses with the plasma membrane, discharging the granular content into the crypt lumen.

The nucleus is basal in position and contains a large nucleolus. Stacks of lamellar cisternae of rough-surfaced endoplasmic reticulum fill much of the cytoplasm around the nucleus. Moderately dense material occupies the cisternae. The Golgi apparatus is seen above the nucleus, and associated with this organelle are vacuoles containing material similar to that comprising the larger secretory granules. Lysosomes (Ly) are common in Paneth cells and have a complex internal structure. Granules, dense droplets, and membranes are found within them. Other structures in the cytoplasm are the mitochondria and free ribosomes. Microvilli occur on the free surface of the cell.

The granules of Paneth cells contain a mucopolysaccharide-protein complex. The secretion of intestinal enzymes, including peptidase, has usually been attributed to these cells, although this function has not been conclusively demonstrated. They are also thought to produce lysozyme, an enzyme that lyses bacteria.

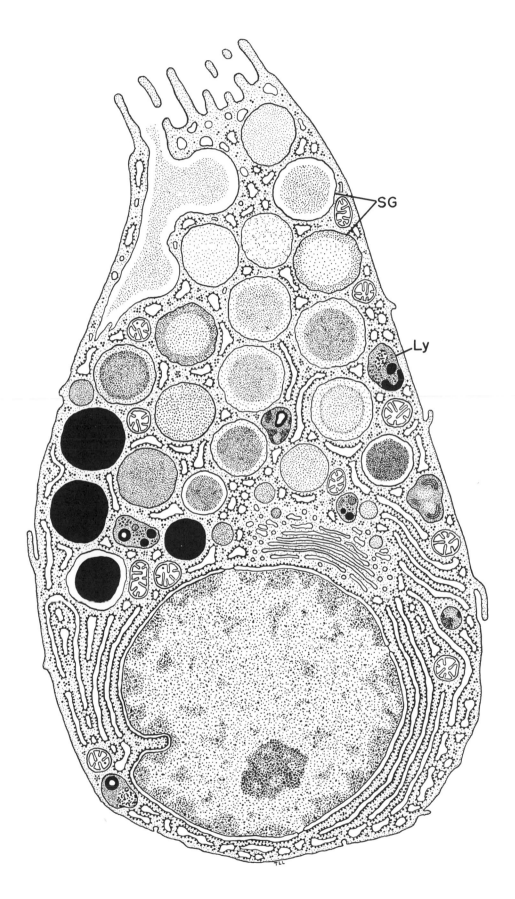

DIGESTIVE SYSTEM

80—INTESTINES: GOBLET CELL

Goblet cells occur singly in the epithelium of the small intestine, especially the ileum, and are very abundant in the large intestine. Their flask or goblet shape is due to the accumulation of mucigen droplets that fill the apical cytoplasm. The nucleus is compressed in the basal region of the cell. Elongated, ribosome-studded cisternae of endoplasmic reticulum, free ribosomes, and mitochondria are found in the small amount of cytoplasmic space around the nucleus, between the granules, and at the periphery of the cell. A Golgi apparatus occurs in a supranuclear position. During elaboration of the secretory droplets the Golgi apparatus is extensive, but it is not as conspicuous as it is in the mature cell illustrated here. The secretory droplets are small in the region of the Golgi apparatus and are larger in the apical regions of the cell. The droplets differ in density, but generally the smaller and less mature granules near the Golgi are denser, while the larger, more mature granules have less density. The apical droplets have a reticular or particulate appearance. Some of these granules appear to fuse, and the membrane enclosing the mucus mass is fused with the plasma membrane. In this manner, the mucus empties into the intestinal lumen. The apical border of the cell is composed of microvilli, and some interdigitations are found laterally. The mucus secretion of the goblet cells lubricates the lumen of the gastrointestinal tract, particularly the large intestine.

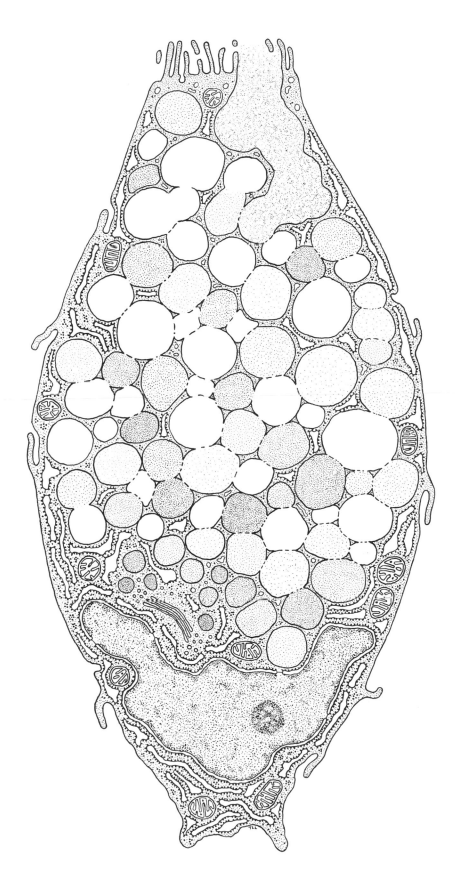

DIGESTIVE SYSTEM

81 — HEPATOCYTE

The hepatic parenchymal cell is polygonal in shape with a large, round nucleus containing a prominent nucleolus. The nuclear envelope is perforated by pores and may be continuous with cisternae of endoplasmic reticulum. Organelles are closely packed in the cytoplasm. Elongated cisternae of rough-surfaced endoplasmic reticulum occur singly or in stacks of two to four. Stretches of membrane separate groups of attached ribosomes. Ribosomes also occur free in the cytoplasm. Tubules of smooth-surfaced endoplasmic reticulum (SER), continuous with rough-surfaced cisternae, form anastomosing networks. The interstices of the smooth-surfaced networks are occupied by glycogen granules. Most of the glycogen granules occur in rosettes or large aggregations referred to as α particles. Single granules, or β particles, also are present. Dense granules occur within some of the cisternae of endoplasmic reticulum (CG).

The Golgi apparatus (G), situated near a bile canaliculus or the nucleus, is composed of parallel lamellae, small vesicles, and larger vacuoles. Some of the Golgi vacuoles contain dense granules that have a diameter of $\sim$500 Å. Elements of the endoplasmic reticulum that face the Golgi region and intervening vesicles contain similar granules. Mitochondria are abundant in liver cells and a few lipid droplets occur. Lysosomes (Ly) are composed of moderately dense, finely granular material and may contain dense granules, opaque material, and remnants of other organelles such as mitochondria. Lysosomes are abundant near bile canaliculi (BC). Another prominent structure in the liver cell is the microbody (Mb), a spherical body with a granular matrix that often contains a denser core or nucleoid. Microbodies contain one or all of the enzymes urate oxidase, catalase, and D-amino acid oxidase. The crystalline core consists of urate oxidase. Man and other animals that lack uricase also lack the dense core within microbodies. Microvilli occur on the cell surfaces limiting bile canaliculi and facing sinusoids (Si). Pinocytotic invaginations occur on the subsinusoidal surface.

The liver cell has been termed a metabolic mill because virtually all the reactions of intermediary metabolism take place in the cell. A major function of the liver is the maintenance of normal blood-glucose levels. Glucose taken up by the cell is stored in the form of glycogen, which can be broken down again to glucose by the action of phosphorylase. Many of the enzymes involved in glycogenesis and glycogenolysis lie free in the hyaloplasm. The smooth-surfaced endoplasmic reticulum, near which the glycogen deposits lie, may be associated with exit of glucose from the cell during glycogenolysis. Glucose-6-phosphatase occurs in the membranes of the reticulum and may be active in transport of glucose across the membrane.

The liver is also responsible for the maintenance of circulating lipid levels. Lipids are synthesized from carbohydrates and proteins in the liver. These, along with lipids in the chylomicrons and those released by adipose cells, are incorporated into lipoproteins, a major vehicle for transport of lipids in the blood. The small dense granules seen in the Golgi region and elsewhere may represent serum lipoproteins that are subsequently released from the cell. Triglycerides are formed from fatty acids in the smooth endoplasmic reticulum. They are combined with protein synthesized by the rough endoplasmic reticulum to form lipoprotein particles. The smooth endoplasmic reticulum may also be the site of cholesterol synthesis.

Synthesis of plasma proteins occurs in the liver. This process presumably takes place on the ribosomes of the rough endoplasmic reticulum. The agranular reticulum contains enzymes that metabolize lipid soluble drugs. Administration of these drugs induces a marked increase and proliferation of smooth endoplasmic reticulum. The structural and enzymatic components of the membranes are synthesized by the rough-surfaced endoplasmic reticulum. Finally, the hepatic cell produces components of bile, including bile acids and bile pigments, and secretes them into the bile canaliculi.

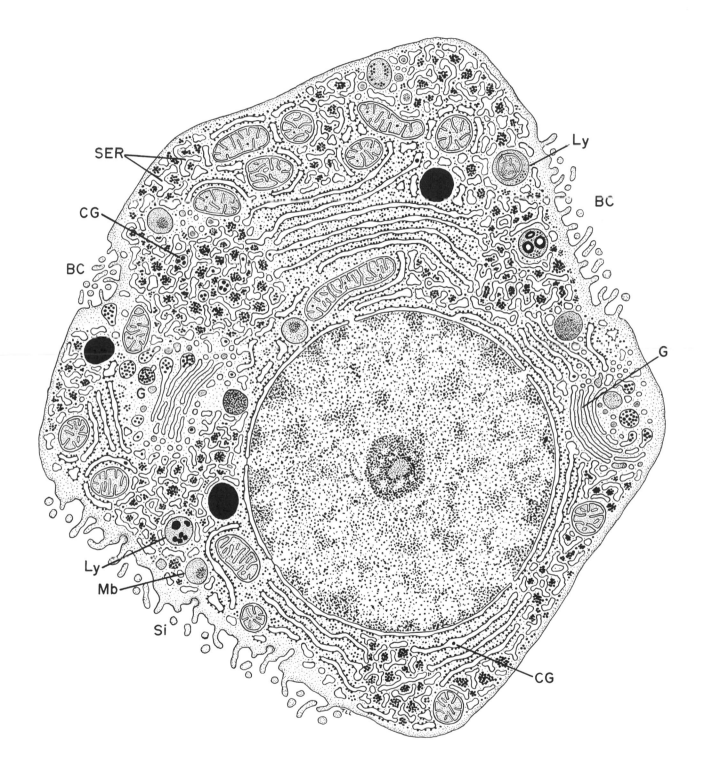

82 — BILE DUCT EPITHELIAL CELL

The bile canaliculi between hepatic cells communicate with interlobular bile ducts lined by flattened or cuboidal cells. The larger intrahepatic bile ducts are lined by a single layer of cuboidal to columnar epithelial cells. The oval nucleus is indented and located toward the base of the cell. Some short stubby microvilli project from the apical surface. A few pits or invaginations are found at the bases of the microvilli. Junctional complexes (not shown) join adjacent cells. Some folds and interdigitations occur on the lateral surfaces, especially near the base of the cell. In the cytoplasm, a Golgi apparatus is found near the nucleus. Spherical to ovoid mitochondria are distributed throughout the cell. A few lysosomes are seen, as well as some short cisternae of rough-surfaced endoplasmic reticulum. Free ribosomes, usually occurring in clusters, are abundant. Bundles of filaments also occur in the cytoplasm.

It has been suggested that the epithelial cells of the duct release a secretory product into the lumen. An extensive capillary system around the ducts might indicate that the duct cells can further modify the bile. Structurally, the cells are relatively unspecialized, and their major function would seem to be to provide channels for the movement of bile from the canaliculi to the gallbladder.

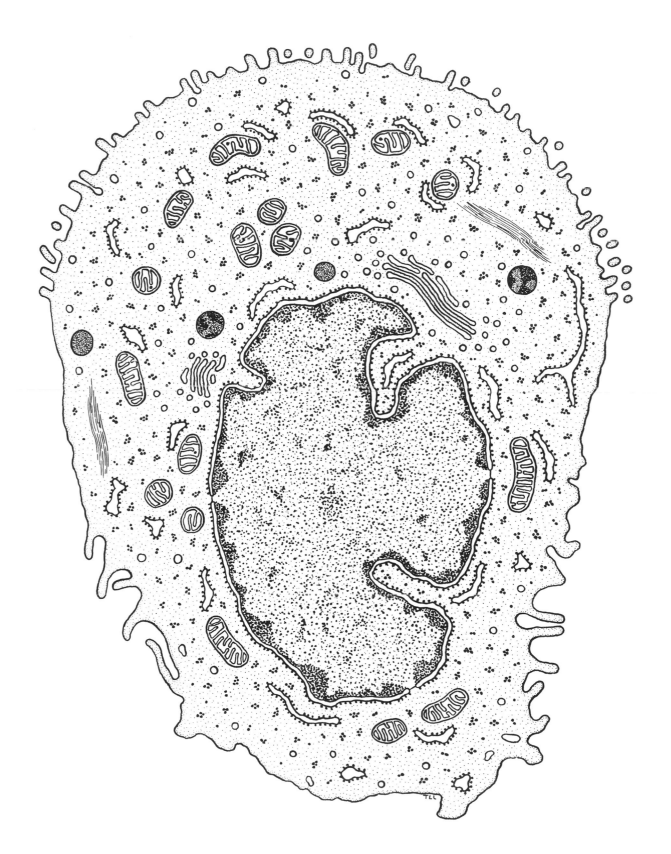

83 — GALLBLADDER EPITHELIAL CELL

The gallbladder is lined by a single layer of tall, club-shaped cells. The apical surface bears short stubby microvilli (Mv) with filamentous material at their tips (originally called *antennulae microvillares*, Yamada). Some apical pits are found at the bases of the microvilli. Junctional complexes (not shown) occur along the apical lateral borders. Infoldings and plications are found on the lateral surface, and these are more complex toward the base of the cell. The oval nucleus, containing a nucleolus, is located in the middle or lower portion of the cell. Organelles are sparse immediately below the apical surface. Round or oval mitochondria occur in the cytoplasm beneath this zone. A small Golgi apparatus is situated near the nucleus. Mitochondria are also numerous in the basal region and often are elongated or filamentous in shape. Some lysosomes, short cisternae of endoplasmic reticulum, and clusters of ribosomes complete the cytoplasmic structures.

The gallbladder serves as a reservoir for bile secreted by the hepatic cells. It also functions to concentrate the bile by withdrawal of water and ions. In gallbladders known to be transporting fluid, the intercellular spaces are distended while apical and basal relationships remain intact. Adenosine triphosphatase (ATPase), which can play a role in active transport of electrolytes, has been detected along the lateral surfaces. Thus, it has been suggested that fluid absorption by the gallbladder takes place by active transport of solute across the lateral cell membranes into the intercellular space (Kaye et al.). This results in an osmotic gradient that causes water to move from the cell and into the intercellular space, distending it. The resultant hydrostatic pressure drives the fluid through the basal lamina into the lamina propria and into proximity with capillaries.

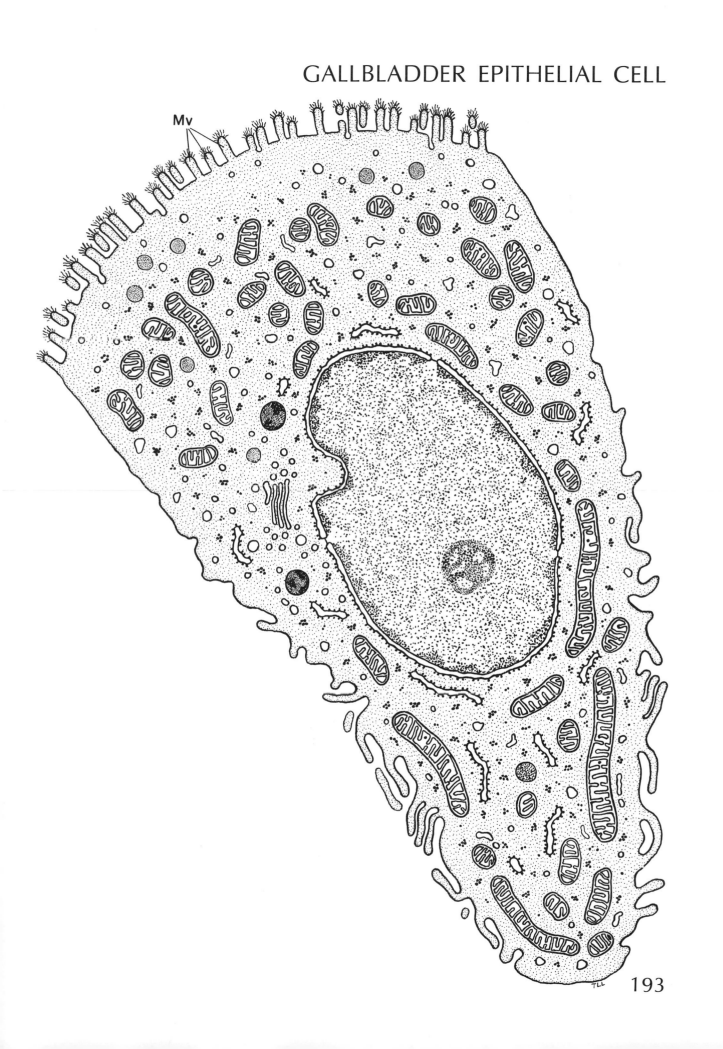

Mv

193

DIGESTIVE SYSTEM

84—PANCREATIC ACINAR CELL

The cells comprising the acini of the exocrine pancreas elaborate digestive enzymes such as trypsin, chymotrypsin, carboxypeptidase, amylase, and lipase. These cells are specialized for protein synthesis. Zymogen granule formation exemplifies the roles of the cellular organelles in the production of a protein for export. The round nucleus is situated in the basal portion of the cell. Finely granular chromatin material is condensed against the inner aspect of the nuclear envelope and occurs around the nucleolus. The nucleolus (Nl) is composed of a fibrous portion and of regions of dense granules that resemble cytoplasmic ribosomes. The nuclear envelope contains many pores and is sometimes in continuity with cisternae of rough-surfaced endoplasmic reticulum. The basal portion of the cell is occupied by the rough-surfaced endoplasmic reticulum (ER), which is organized into an elaborate system of parallel and intercommunicating lamellae. Sometimes, granules with a density similar to that of zymogen granules are found within the cisternae of the endoplasmic reticulum and are called cisternal granules (CG). Ribosomes also occur free in the cytoplasm. Mitochondria (M) are most numerous basally and contain many transverse cristae and opaque granules in the matrix. Cisternae of endoplasmic reticulum sometimes envelop the mitochondria. An elaborate Golgi apparatus (G) is situated above the nucleus and consists of stacks of membranous lamellae, small vesicles, and larger vacuoles. Some Golgi vacuoles or condensing vacuoles (CV) contain moderately dense material and represent transitional stages in zymogen granule formation. Occasionally, smooth-surfaced outpocketings occur on the surfaces of the rough-surfaced endoplasmic reticulum that face the Golgi apparatus.

Zymogen granules (ZG) are abundant in the apical cytoplasm. These structures, containing the inactive precursors of the pancreatic enzymes, are composed of dense material and are bounded by a membrane. Discharge of the granules occurs by fusion of the membrane bounding the granule with the apical plasma membrane. Several granules sometimes fuse and form a series of interconnected granules. Dense material also occurs in the lumen. A few microvilli extend from the apical surface. The surface plasma membrane is coated with a feltwork of fibrillar material. Lysosomes, multivesicular bodies, and centrioles have been observed in acinar cells.

The cytological pathway of protein synthesis was first determined in the pancreatic acinar cell (Palade, Siekevitz, Jamieson). Incorporation of amino acids into proteins, the digestive enzymes or zymogens, takes place on the ribosomes attached to the endoplasmic reticulum. The proteins are transferred to the cisternae of the endoplasmic reticulum. They are then transported to the Golgi complex within small vesicles. Here, the secretory product is concentrated within condensing vacuoles. These give rise to mature zymogen granules that migrate and accumulate at the apex of the cell. The proteins are then discharged into the acinar lumen. Many cells have in common the synthesis of secretory proteins, segregation and temporary storage of the secretory products, and discharge of the products into a glandular lumen upon appropriate stimulation. It is thought that the mechanism by which these processes are carried out in other cells is similar to that elucidated for the pancreas.

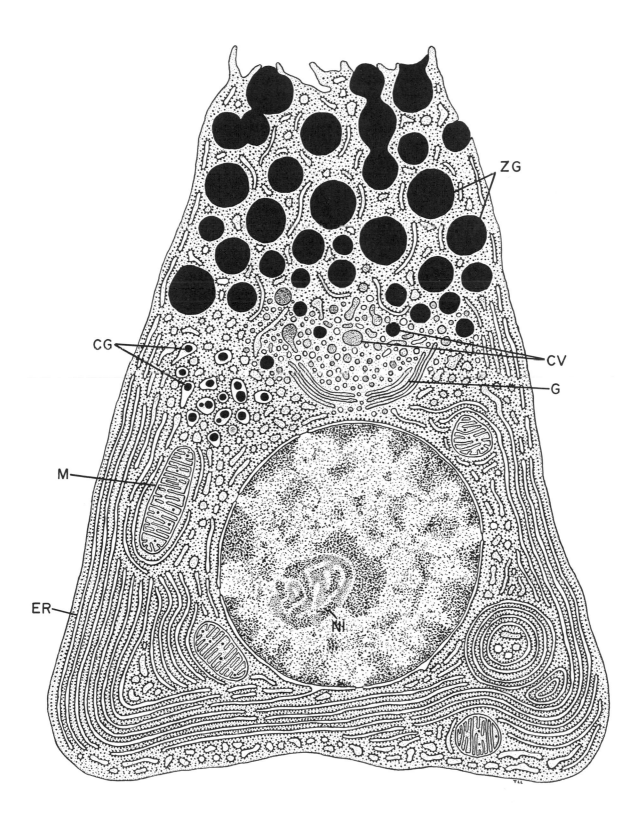

85—PANCREATIC DUCT EPITHELIAL CELL

The smallest ducts of the pancreas arise in the acini. The duct cells are enclosed by the acinar cells and are known as centroacinar cells. These small ducts drain into the intralobular or intercalated ducts, which open into the larger interlobular ducts. The cells gradually change in shape from flat to cuboidal to columnar, progressing from the smaller to the larger ducts, but all are similar in cytoplasmic structure. The luminal border bears some short microvilli. Infoldings and interdigitations are found laterally, and apically there are junctional complexes (not shown). The cells have a large central nucleus. A small Golgi apparatus is situated near the nucleus. A moderate number of mitochondria and some lysosomes are found in the cytoplasm. Cisternae of endoplasmic reticulum and clusters of ribosomes occur in small numbers. Bundles of cytoplasmic filaments are fairly common. The cells are similar to those in other conducting ducts. It has been suggested that the duct cells secrete electrolytes, and carbonic anhydrase has been identified within these cells.

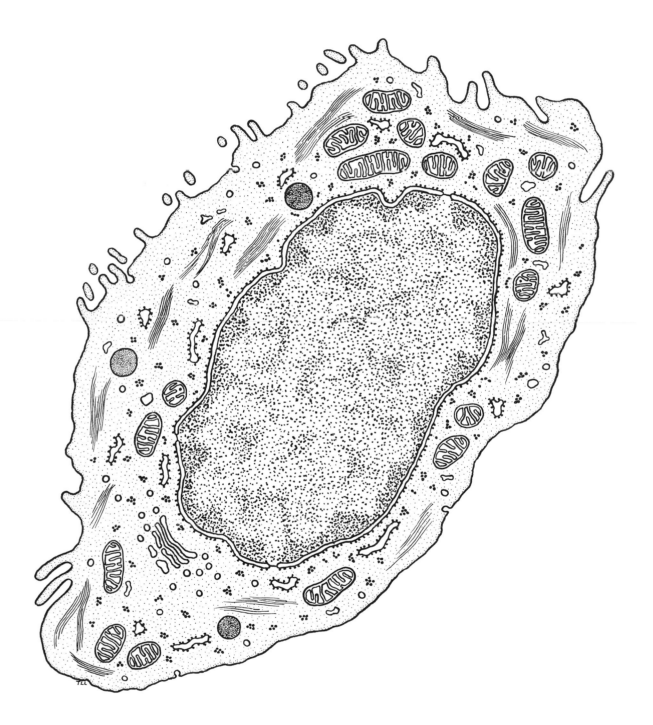

REFERENCES

Amsterdam, A., I. Ohad, and M. Schramm, 1969. Dynamic changes in the ultrastructure of the acinar cell of the rat parotid gland during the secretory cycle. J. Cell Biol., 41:753-773.

Behnke, O. and H. Moe, 1964. An electron microscope study of mature and differentiating Paneth cells in the rat, especially of their endoplasmic reticulum and lysosomes. J. Cell Biol., 22:633-652.

Caramia, F., 1966. Ultrastructure of mouse submaxillary gland. I. Sexual differences. J. Ultrastruct. Res., 16:505-523.

Cardell, R. R., S. Badenhausen, and K. R. Porter, 1967. Intestinal triglyceride absorption in the rat. An electron microscopical study. J. Cell Biol., 34:123-156.

Corpron, R. E., 1966. The ultrastructure of the gastric mucosa in normal and hypophysectomized rats. Amer. J. Anat., 118:53-90.

Ekholm, R., T. Zelander, and Y. Edlund, 1962. The ultrastructural organization of the rat exocrine pancreas. I. Acinar cells. J. Ultrastruct. Res., 7:61-72.

Ekholm, R., T. Zelander, and Y. Edlund, 1962. The ultrastructural organization of the rat exocrine pancreas. II. Centroacinar cells, intercalary and intralobular ducts. J. Ultrastruct. Res., 7:73-83.

Farquhar, M. G., and G. E. Palade, 1963. Junctional complexes in various epithelia. J. Cell Biol., 17:375-412.

Forssmann, W. G., L. Orci, R. Pictet, A. E. Renold, and C. Rouiller, 1969. The endocrine cells in the epithelium of the gastrointestinal mucosa of the rat. An electron microscopic study. J. Cell Biol., 40:692-715.

Freeman, J. A., 1966. Goblet cell fine structure. Anat. Rec., 154:121-148.

Friend, D. S., 1965. The fine structure of Brunner's glands in the mouse. J. Cell Biol., 25:563-576.

Garant, P. R. and J. Nalbandian, 1968. Observations on the ultrastructure of ameloblasts with special reference to the Golgi complex and related components. J. Ultrastruct. Res., 23:427-443.

Garant, P. R., G. Szabo, and J. Nalbandian, 1968. The fine structure of the mouse odontoblast. Arch. Oral Biol., 13:857-876.

Ichikawa, A., 1965. Fine structural changes in response to hormonal stimulation of the perfused canine pancreas. J. Cell Biol., 24:369-385.

Ito, S. and R. J. Winchester, 1963. The fine structure of the gastric mucosa in the bat. J. Cell Biol., 16:541-578.

Jamieson, J. D. and G. E. Palade, 1967. Intracellular transport of secretory proteins in the pancreatic exocrine cell. I. Role of the peripheral elements of the Golgi complex. J. Cell Biol., 34:577-596.

Jamieson, J. D. and G. E. Palade, 1967. Intracellular transport of secretory proteins in the pancreatic exocrine cell. II. Transport to condensing vacuoles and zymogen granules. J. Cell Biol., 34:597-615.

Kaye, G. I., H. O. Wheeler, R. T. Whitlock, and N. Lane, 1966. Fluid transport in the rabbit gallbladder. A combined physiological and electron microscope study. J. Cell Biol., 30:237-268.

Palade, G. E., P. Siekevitz, and L. H. Caro, 1962. Structure, chemistry, and function of the pancreatic exocrine cell. In de Reuck, A. V. S. and M. P. Cameron (eds.), The Exocrine Pancreas. Little, Brown, and Co., Boston, 23-49.

Scalzi, H. A., 1967. The cytoarchitecture of gustatory receptors from the rabbit foliate papillae. Z. Zellforsch., 80:413-435.

Sedar, A. W. and M. H. F. Friedman, 1961. Correlation of the fine structure of the gastric parietal cell (dog) with functional activity of the stomach. J. Biophys. Biochem. Cytol., 11:349-364.

Senior, J. R., 1964. Intestinal absorption of fats. J. Lipid Res., 5:495-521.

Sjöstrand, T. S., 1962. The fine structure of the exocrine pancreas cells. In de Reuck, A. V. S. and M. P. Cameron (eds.), The Exocrine Pancreas. Little, Brown and Co., Boston, 1-19.

Staley, M. W. and J. S. Trier, 1965. Morphologic heterogeneity of mouse Paneth cell granules before and after secretory stimulation. Amer. J. Anat., 117:365-384.

Trier, J. S., 1963. Studies on small intestinal crypt epithelium. I. The fine structure of the crypt epithelium of the proximal small intestine of fasting humans. J. Cell Biol., 18:599-620.

Wilborn, W. H. and J. M. Shackleford, 1969. The cytology of submandibular glands of the opossum. J. Morph., 128:1-34.

Yamada, K., 1968. Aspects of the fine structure of the intrahepatic bile duct epithelium in normal and cholecystectomized mice. J. Morph., 124:1-22.

RESPIRATORY SYSTEM

RESPIRATORY SYSTEM

86—OLFACTORY EPITHELIUM: OLFACTORY CELL

The pseudostratified epithelium of the olfactory mucosa consists of three cell types: olfactory cells, supporting or sustentacular cells, and basal cells.

The olfactory cells are bipolar ganglion cells with their cell bodies located in the epithelium; a distal process terminating in an olfactory vesicle (OV), which extends above the free surface; and a proximal axonal process (Ax), which terminates in the olfactory bulb.

The nucleus is oval in shape and has masses of condensed chromatin. A large Golgi apparatus is found in the perikaryon in a supranuclear position. Many small vesicles are associated with the Golgi apparatus. A few flattened cisternae of rough-surfaced endoplasmic reticulum, as well as smooth membranes, are found in the cytoplasm. Glycogen granules are widely distributed. Free ribosomes are common, and a few dense bodies or lysosomes are seen. Mitochondria in the perikaryon are round or oval, with straight or curved cristae.

The distal process or olfactory rod extends upward from the perikaryon, ending in the dilated olfactory vesicle (OV). Microtubules (Mt) are found in large numbers in the process, and some smooth channels of endoplasmic reticulum occur as well. The mitochondria are elongated in the direction of the process, and the cristae are also arranged longitudinally. The olfactory vesicle is a bulb-shaped expansion projecting slightly above the epithelial surface. Cilia (C) (10 to 20 per cell) project in all directions from the vesicle and are embedded in the thick mucous layer coating the epithelium. The cilia turn to run parallel to and a short distance below the surface. In their proximal regions, the cilia have the usual nine-plus-two arrangement of tubules, but proceeding distally, fewer tubules occur. At the same time, the diameter of the cilia is reduced until they are about the diameter of villi. The olfactory vesicle itself contains the basal bodies (BB) of the cilia, microtubules projecting up from the process, rounded mitochondria, and some vesicles and smooth membranous channels. A junctional specialization is formed between the proximal portion of the olfactory vesicle and the apical lateral borders of adjacent supporting cells.

The axonal nerve fiber (Ax) extends from the opposite pole of the perikaryon and penetrates the basement membrane of the epithelium. The nerve processes of nearby cells combine into bundles that join others to form the olfactory nerve. The process is thin and contains some elongated mitochondria and microtubules (Mt) (neurotubules).

The receptor cell of the olfactory epithelium is a ciliated nerve cell. The cilia themselves may serve as the initial receptor sites for olfactory stimuli and also may be capable of limited movement. Presumably, volatile, odoriferous substances, which can be detected in extremely low concentrations, become dissolved in the fluid coating of the epithelium and then combine with receptor sites on the slender cilia. There are thought to be 7 to 10 primary odors and, hence, receptor sites. Sensations of various odors result from different combinations of these primary odors.

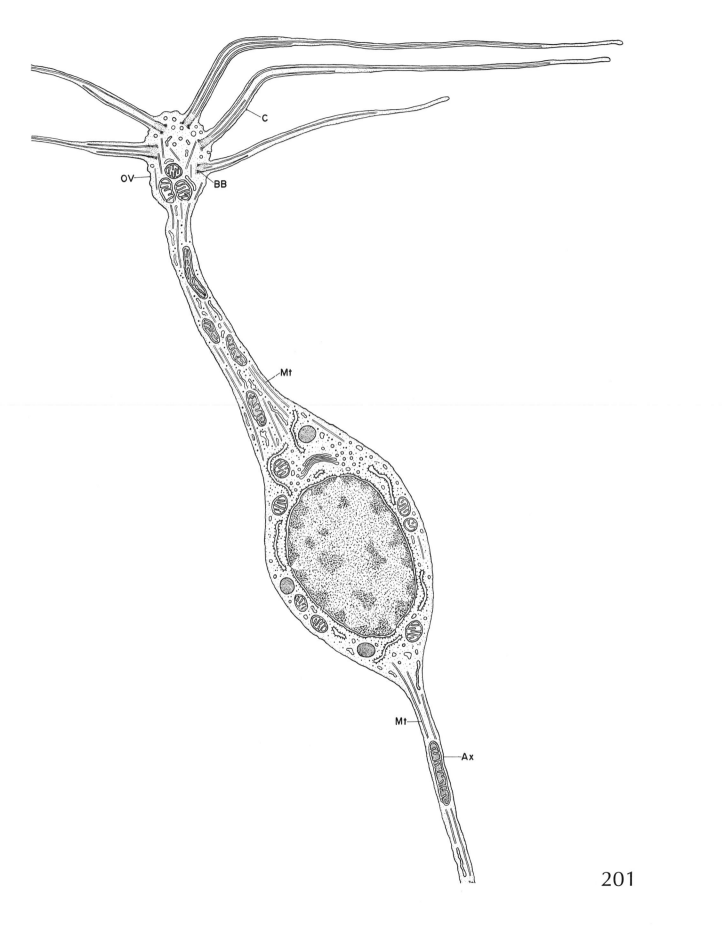

RESPIRATORY SYSTEM

87 — OLFACTORY EPITHELIUM: SUSTENTACULAR CELL

The supporting or sustentacular cells of the olfactory epithelium are high-columnar and border the free surface. The surface of the cell is provided with long microvilli that mingle with the ciliary extensions of the olfactory cells. Immediately below the surface is a terminal web area that is relatively free of organelles. The apical region of the cell is occupied by mucous granules (MG), smooth membranes, and mitochondria. The mucous granules are irregular in shape, and most are relatively dense. A Golgi apparatus from which the granules originate is located above the nucleus. The smooth endoplasmic reticulum (SER) consists of tubules that run roughly parallel to the long axis of the cell. They branch and anastomose to some extent and envelop the mitochondria. Rough-surfaced cisternae are short and much less abundant. The mitochondria are also arranged longitudinally, and they have lamellar cristae.

The nucleus is located in the upper half of the cell and is at a higher level than that of the olfactory cell. The chromatin is dispersed for the most part. The smooth endoplasmic reticulum sometimes forms loose whorls in the basal cytoplasm. Glycogen granules are common in this area as well as dense bodies resembling lysosomes and lipofuscin pigment bodies. Slender foot processes (FP) occur at the base of the cell and may loosely enclose bundles of nerve fibers. Filaments and microtubules extend into the cytoplasmic processes.

The sustentacular cells probably contribute to or modify the secretory product of Bowman's glands that coats the epithelial surface. The chemical nature of these secretions is undefined, but one component appears to be mucus. These cells have some cytological resemblance — particularly, the extensive smooth endoplasmic reticulum — to the dark cells of Bowman's gland, and their secretory products could be similar as well.

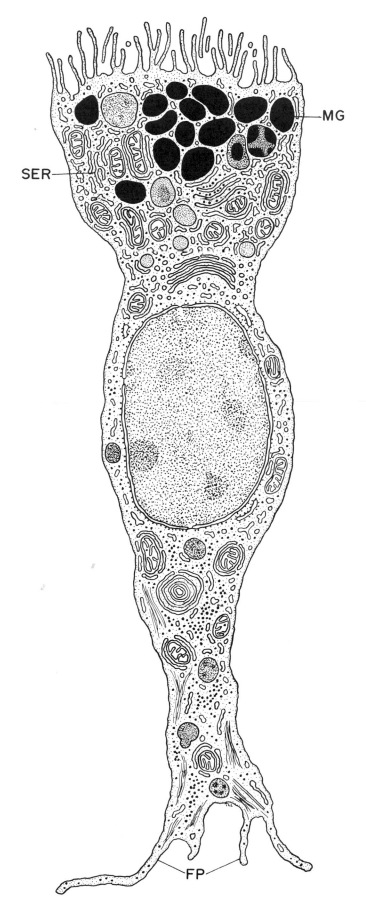

88—OLFACTORY EPITHELIUM: BASAL CELL

The basal cells are found at the bottom of the epithelium adjacent to the basement lamina. The cells are rounded and have a central, indented nucleus. Cytoplasmic processes extend from the cell body and invest bundles of olfactory cell axons (Ax), or interdigitate with processes of adjacent cells. The relatively scant cytoplasm contains a small Golgi apparatus, a few rough-surfaced cisternae, ribosomes, large mitochondria, and lysosomes. Glycogen granules are common, and bundles of filaments run in the cytoplasm. These cells appear relatively unspecialized and may be capable of giving rise to other cells in the epithelium.

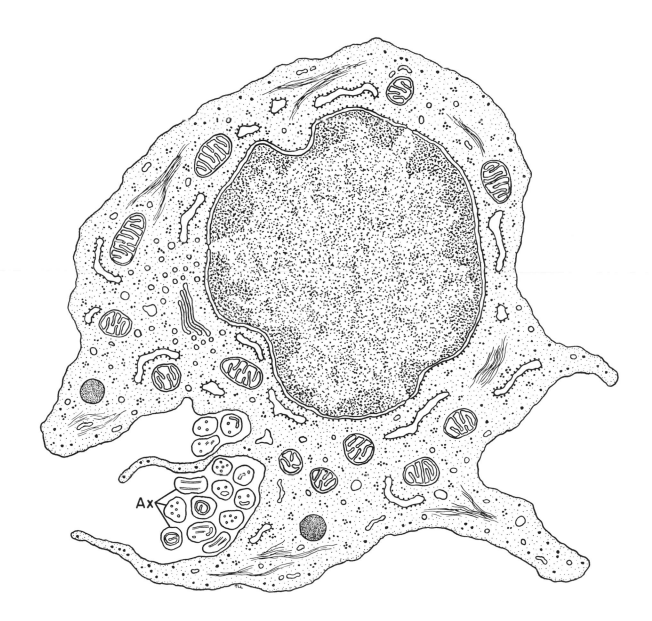

89 — BOWMAN'S GLAND: LIGHT CELL

Bowman's or olfactory glands are branched tubuloalveolar glands located in the connective tissue beneath the epithelium and opening to the surface by way of ducts. Two cell types, termed light and dark on the basis of the appearance of their cytoplasm at low magnification, comprise the Bowman's gland. Both cell types are pyramidal and border the lumen of the gland.

Light cells are characterized by an extensive rough-surfaced endoplasmic reticulum (ER) that fills most of the cell. The cisternae are dilated and contain moderately dense material. A Golgi apparatus (G) occurs above the nucleus and has vesicles and small dense granules associated with it. Vesicles and granules also accumulate in the apex of the cell from which short microvilli (Mv) extend into the lumen. Elongated mitochondria and ribosomes occupy the cytoplasm between cisternae. The basally situated nucleus contains clumps of condensed heterochromatin. Bowman's glands constantly discharge a serous fluid to the surface, providing a fresh supply of solvent for odoriferous substances and keeping the surface of the epithelium moist. The light cell of the gland is clearly active in protein synthesis, but the exact nature of its product is unknown.

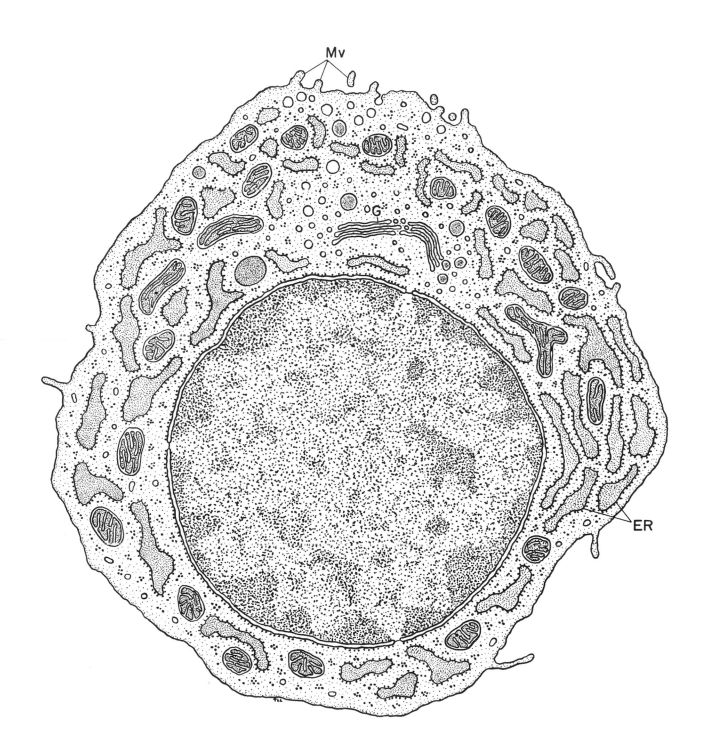

90 — BOWMAN'S GLAND: DARK CELL

Dark cells of Bowman's gland, like the light cells, are pyramidal with their apical surfaces bearing microvilli and bordering the lumen. Some elongated villous processes also protrude from the lateral and basal surfaces. The most conspicuous features of these cells are the secretory granules and the agranular endoplasmic reticulum. The secretory granules (SG) are spherical and composed of material of low density. The largest granules accumulate at the apex of the cell, but smaller vacuoles, usually containing less-dense material, are found in the vicinity of the Golgi apparatus. This organelle is situated above the nucleus. Much of the cytoplasm is occupied by tubules of smooth-surfaced endoplasmic reticulum (SER). The cisternae tend to run in parallel, and in some places they are tightly packed in concentric arrangements to form whorl systems. Rough-surfaced cisternae are few and are often continuous with the smooth tubules. Others occur as isolated cisterns at the base of the cell. Mitochondria are large, and their cristae are mostly lamellar and may be transverse or longitudinal. The mitochondria are closely invested by the smooth tubules. A small number of ribosomes occur free in the cytoplasm.

Other cells with extensive agranular endoplasmic reticulum are engaged in steroid or lipid synthesis. The dark cells of Bowman's gland, as well as the sustentacular cells of the epithelium that show a similar specialization, might add a lipid component to the fluid covering the olfactory epithelium. In this case, a mechanism would be available for concentration of odoriferous substances, since such substances are much more soluble in lipids than in water.

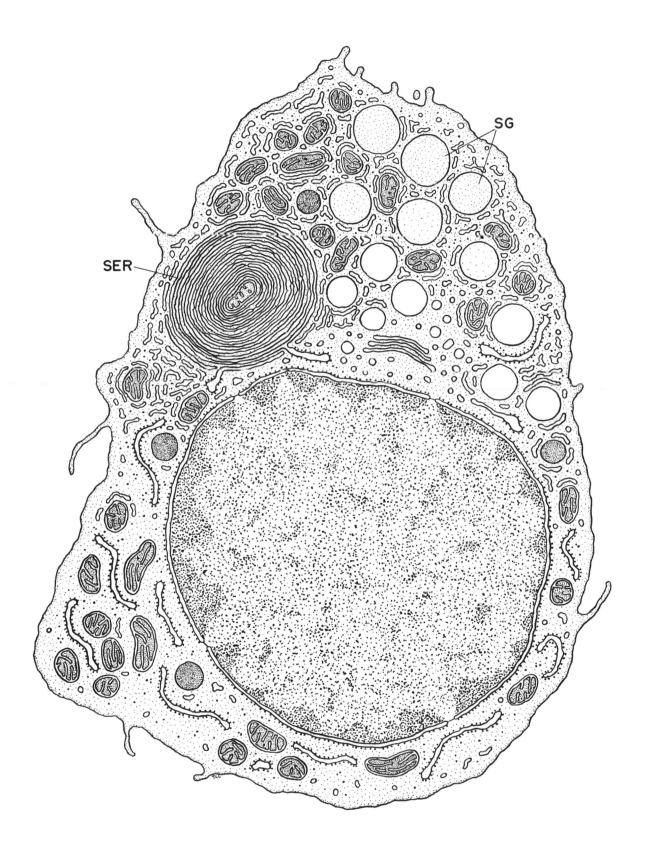

RESPIRATORY SYSTEM

91—RESPIRATORY EPITHELIUM: CILIATED EPITHELIAL CELL

The greater part of the respiratory tract is lined by ciliated pseudostratified columnar epithelium. Four cell types comprise this epithelium: ciliated cells, mucous cells, brush cells, and basal cells. All the cells rest on the basement membrane, and all except the basal cells extend to the surface. The cell nuclei, however, lie at different levels, so that the epithelium appears to be stratified.

The ciliated cells are the most numerous. The nucleus is situated in the narrow basal region of the cell. A Golgi apparatus occurs above the nucleus, and a system of vesicles and short membranous profiles constitute the smooth-surfaced endoplasmic reticulum, which is more extensive in the apical region of the cell. Mitochondria are also most common in this area, and some lysosomes are found in the cytoplasm. Only a few short cisternae of rough-surfaced endoplasmic reticulum are present, but clusters of free ribosomes are abundant. The ciliated cells contain filaments (Fl), especially below the surface in a zone largely devoid of other organelles. The hyaloplasm is of relatively low density.

The apical surface of the cell is highly specialized with cilia (C) and microvilli. The microvilli are short and project between the cilia. The cilia are provided with an internal system of fibrils arranged with nine double fibrils surrounding a central pair. The outer fibrils extend into a basal body (BB) below the cell surface. In cross section (upper right of Plate), the basal body is seen as composed of nine triplets, identical to a centriole. From the basal body, a striated rootlet (Rt) extends deeper into the cytoplasm. The action of the cilia is responsible for removal of foreign particles, such as bacteria, from the respiratory tract.

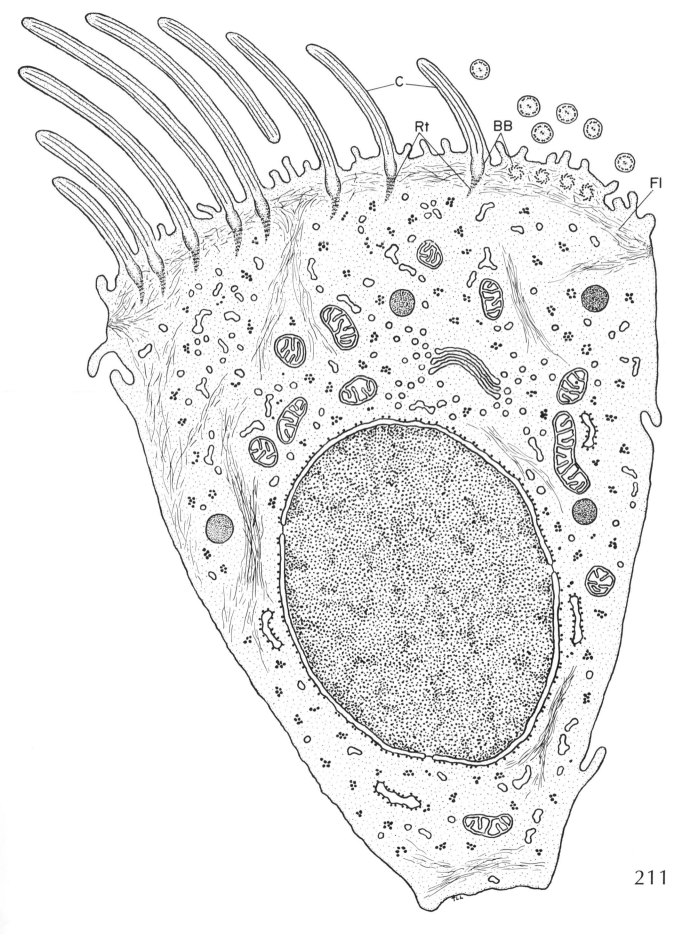

C

Rt

BB

Fl

RESPIRATORY SYSTEM

92—RESPIRATORY EPITHELIUM: MUCOUS CELL

The mucous or goblet cell of the respiratory epithelium is similar to mucus-producing cells found elsewhere. They occur singly between the other cell types of the epithelium. These cells are columnar with the nucleus in the basal region and the secretory granules filling the apical portion. The secretory granules coalesce and fuse with the surface membrane. Some elongated cisternae of rough-surfaced endoplasmic reticulum are found in the lower portion of the cell. Free ribosomes are common. An extensive Golgi apparatus occurs above the nucleus and has immature secretory granules associated with it. Some mitochondria are found throughout the cytoplasm. Short microvilli extend from the free surface. The mucous secretion of these cells, along with trapped particles, is swept toward the pharynx by the action of the ciliated cells.

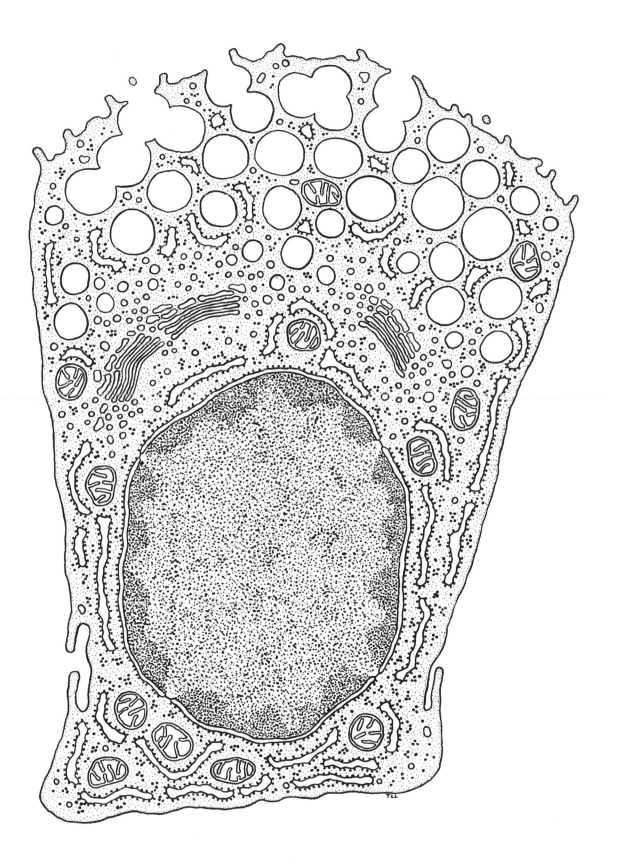

93 — RESPIRATORY EPITHELIUM: BRUSH CELL

The third cell type in the respiratory epithelium is the brush cell. It has also been observed in the alveolar lining. In the large airways, the cell is columnar with the nucleus toward the base. It is characterized by short microvilli (Mv) on the free surface. Bundles of filaments (Fl) occur in the core of the villi and extend downward to join larger bundles that abound in the cytoplasm. Small vesicles occur below the cell's free surface. Mitochondria are most numerous in the apical region and tend to be elongated and run parallel to the long axis of the cell. Smooth-surfaced endoplasmic reticulum occurs in the same area, while rough-surfaced cisternae are more common basally. Free ribosomes are found, and glycogen granules are abundant. A small Golgi complex occurs. The function of this cell is not definitely known. It has been suggested that the cell is chemoreceptive, absorptive, or a transitional stage in formation of a ciliated cell, or that it represents a phase of goblet cell activity.

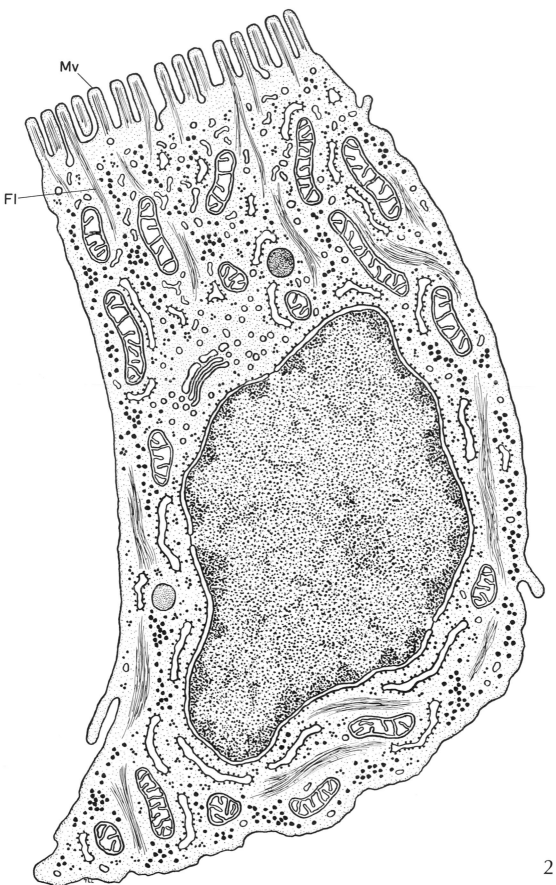

Mv

Fl

RESPIRATORY SYSTEM

94 — RESPIRATORY EPITHELIUM: BASAL CELL

The polygonal or triangular basal cells are the fourth cell type in the epithelium. They form a layer next to the basement membrane underlying the tracheal epithelium. The nucleus, usually containing a nucleolus, occupies a large portion of the cell. Organelles are relatively sparse in the cytoplasm and include some mitochondria, lysosomes, and short cisternae of rough-surfaced endoplasmic reticulum. A few free ribosomes are found. Bundles of tonofilaments are abundant, and some terminate on the small desmosomes found where lateral surfaces of basal cells contact one another. The basal cells are thought to form a germinal layer that gives rise to other cell types of the epithelium. In their uncomplicated structure they are similar to basal cells occurring at the base of other epithelia.

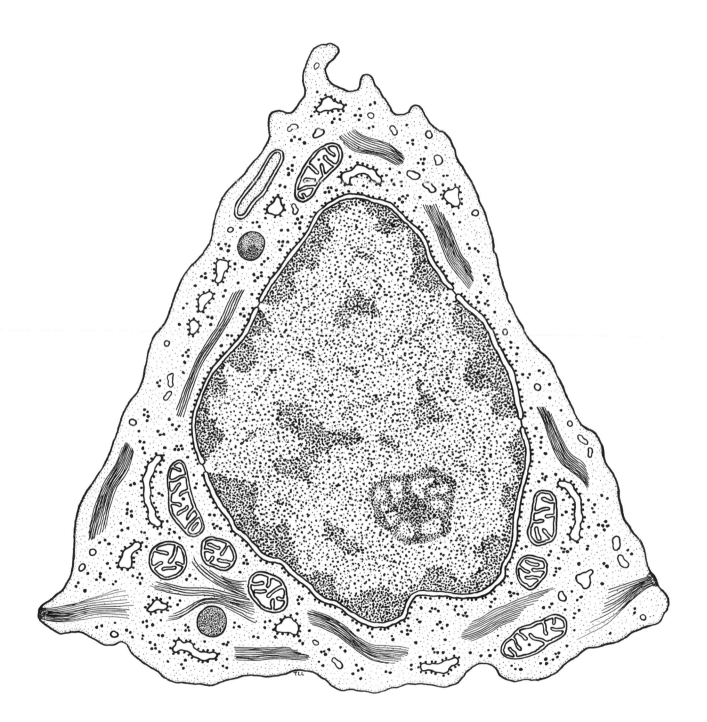

RESPIRATORY SYSTEM

95—LUNG: PULMONARY EPITHELIAL CELL

Pulmonary epithelial cells (small alveolar cells) are flattened cells lining the alveoli. Only the nuclear region of the cell projects into the alveolar space (AS). The nucleus is flattened, and the perinuclear cytoplasm is narrow and contains a sparse complement of organelles. A small Golgi apparatus, small uncomplicated mitochondria, an occasional lysosome, and some short cisternae of rough-surfaced endoplasmic reticulum are found. The epithelial cells also contain small vesicles that are apparently pinocytotic in nature, since some are associated with the cell surfaces.

The remainder of the cell extends over large distances and is attenuated but not interrupted. The capillaries of the alveoli are of the continuous type and contain pinocytotic vesicles (CL, capillary lumen). In some places the endothelium is extremely thin, but it is not fenestrated. The basement laminae (BL) of the epithelial cell and endothelial cell fuse where the the cells are opposed. Thus, the entire barrier over which gaseous exchange between air and blood occurs is very thin (as thin as 100 mμ) and consists of epithelial cell cytoplasm, basement lamina, and endothelial cell cytoplasm. Oxygen and carbon dioxide are thought to pass over this barrier by passive diffusion.

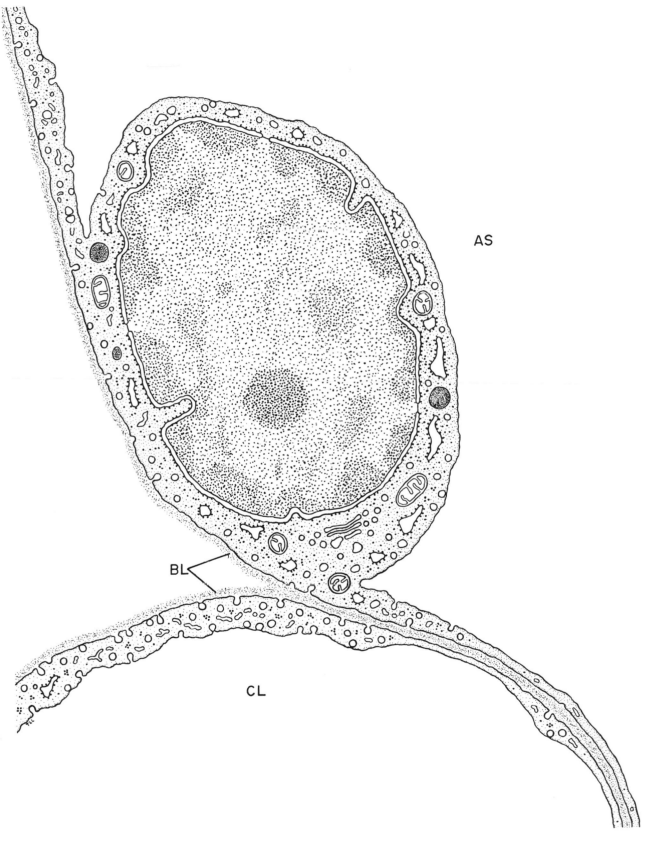

AS

BL

CL

96—LUNG: GREAT ALVEOLAR CELL (SEPTAL CELL)

Great alveolar cells (septal cells, granular pneumocytes) are roughly cuboidal with a central nucleus. They rest on a basement lamina and send short microvilli into the alveolar space. A characteristic feature of the cell is the presence of spherical or ovoid membrane-bounded inclusions termed cytosomes (Cy), or multilamellar bodies, most abundant in the apical portion of the cell. These structures have a heterogeneous content, most containing concentric, osmiophilic lamellae in an electron-lucent matrix. Some also contain vesicles and moderately dense amorphous substance. Some of the cytosomes are in close apposition to the surface plasma membrane; some are even fused with it, so that the contents are exposed to the alveolar space.

The rough-surfaced endoplasmic reticulum is extensive and consists of relatively short, dilated cisternae located throughout the cytoplasm. Free ribosomes are also found. The Golgi apparatus is large and is usually located above the nucleus. It consists of a stack of lamellae and many small vesicles. Mitochondria are numerous and well-developed with closely packed cristae and a dense matrix. Multivesicular bodies (MvB) occur in large numbers. Differences in the structure of multivesicular bodies suggest the formation of cytosomes. Some contain vesicles lying in a clear matrix. In others, the intervesicular matrix is dense and contains a few dense lamellae. In larger bodies, the lamellated material is predominant and the vesicles are restricted to the periphery.

The cytosomes have a high content of lipid, especially phospholipid. The lipid appears to be a secretory product of the cell that is released into the alveolar space. It has been suggested that the fatty material coats the surface of the alveolar epithelium and acts as a surfactant that lowers alveolar surface tension. In the absence of this material, the surface tension would be sufficient to collapse the alveoli.

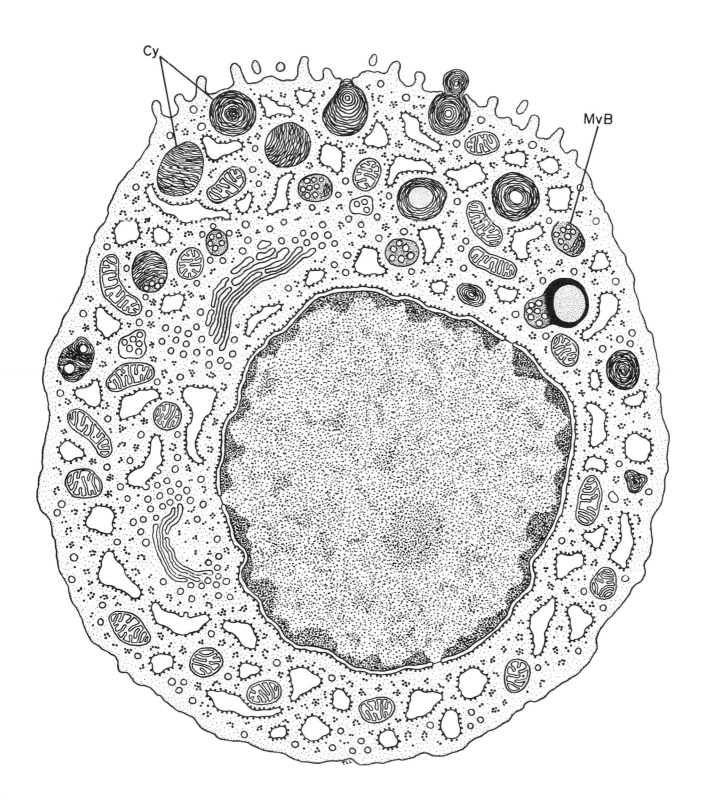

97—LUNG: ALVEOLAR MACROPHAGE

Bordering the lumen of the alveolus are wandering cells known as alveolar phagocytes or macrophages. These cells are rounded, with surface villi or pseudopodia. The nucleus is central in position. In cytoplasmic structure, the cells are somewhat similar to macrophages. Most notable are a number of variably shaped dense granules that probably correspond to lysosomes (Ly). Some are large and contain particulate material within the dense matrix. The cells also contain a few vacuoles with a lamellar content; these vacuoles closely resemble the cytosomes of the great alveolar cells. A Golgi apparatus occurs near the nucleus. There are free ribosomes and some short cisternae of rough-surfaced endoplasmic reticulum. Mitochondria are distributed in the cytoplasm. There is a cortical, ectoplasmic zone that contains few organelles except for an occasional vesicle.

These cells are thought to phagocytose dust and other inhaled particles. For this reason, they have been called dust cells. Their origin, however, is not very clear. They may be derived from hematogenous monocytes and thus represent true macrophages. On the other hand, it has been postulated that they originate through transformation of great alveolar cells that have become detached from the alveolar wall.

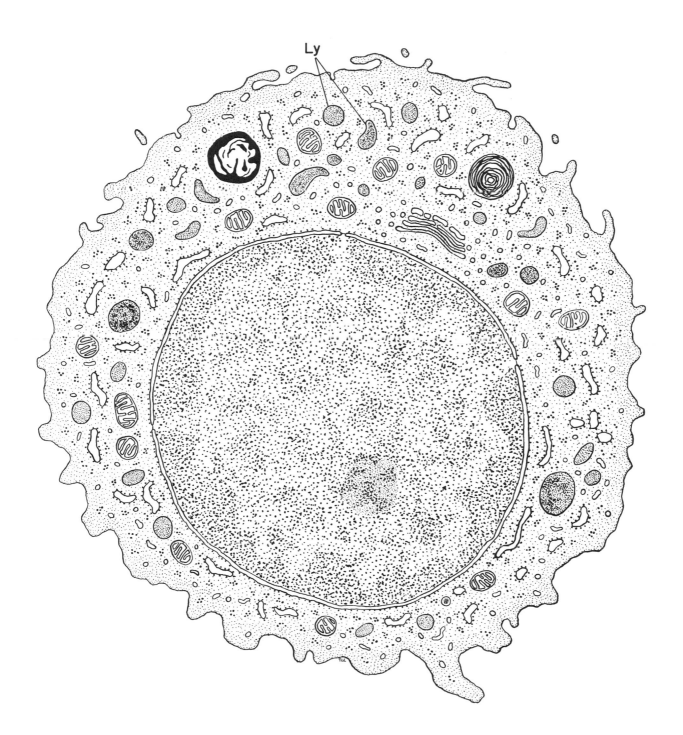

Ly

REFERENCES

Frisch, D., 1967. Ultrastructure of mouse olfactory mucosa. Amer. J. Anat., 121:87-120.

Kistler, G. S., P. R. B. Caldwell, and E. R. Weibel, 1967. Development of fine structural damage to alveolar and capillary lining cells in oxygen-poisoned rat lungs. J. Cell Biol., 32:605-628.

Meyrick, B. and L. Reid, 1968. The alveolar brush cell in rat lung—a third pneumonocyte. J. Ultrastruct. Res., 23:71-80.

Rhodin, J. A. G., 1966. Ultrastructure and function of the human tracheal mucosa. Amer. Rev. Resp. Dis., 93:1-15.

Sorokin, S. P., 1967. A morphologic and cytochemical study on the great alveolar cell. J. Histochem. Cytochem., 14:884-897.

URINARY SYSTEM

98 — GLOMERULAR EPITHELIAL CELL AND CAPILLARY

The glomerular or visceral epithelium of the kidney envelops the capillaries of the glomerulus. The epithelium is composed of cells, called podocytes, which have a central nucleus and several processes radiating from the perikaryon. The major processes give rise to many small secondary foot processes (FP), or pedicels.

The perikaryon of the podocytes contains one or more well-developed Golgi complexes, mitochondria, and rough-surfaced cisternae of endoplasmic reticulum. The latter may be elongated and narrow but more often are short and dilated. Ribosomes also occur free and, often, in clusters in the hyaloplasm. Many small vesicles and vacuoles are present in association with the Golgi complexes and throughout the cytoplasm. Filaments (Fl) and microtubules are abundant in the perikaryon and extend into the pedicels.

The foot processes extend to the basement lamina (BL). They are slightly dilated at their ends and separated from their neighbors by narrow gaps called filtration slits (FiS). The gaps between the foot processes appear to be bridged by a thin diaphragm or slit membrane. The relatively thick basement lamina forms a continuous layer interposed between the foot processes and endothelial cells of the capillaries. It is composed of fine filaments embedded in a homogeneous matrix. The endothelial cells of the capillaries are extremely attenuated, except where the nucleus and surrounding cytoplasm project into the capillary lumen (CL). The thin endothelial layer is perforated by fenestrae (Fen) or pores. As a result, the blood plasma is in direct contact with the basement lamina, although some investigators believe a diaphragm extends across the pores. The basement lamina and slit membranes form barriers between the capillary lumen and the urinary space. A portion of an erythrocyte is illustrated in the capillary lumen.

Fluid from the glomerular capillaries is filtered into the surrounding space of Bowman's capsule. It must, therefore, pass through the pores in the endothelium, through the basement lamina, and through the filtration slits between foot processes of the podocytes. The fluid in Bowman's capsule is an ultrafiltrate of plasma containing small molecules but free of large protein molecules. Extremely large molecules, such as ferritin, are impeded by the basement lamina. Smaller molecules, though with a molecular weight of over 45,000, can pass through the basement lamina but are held up at the level of the filtration slits, which may represent the principal filtration barrier. Smaller molecules can pass into the urinary space.

Another cell type is interposed between the basement lamina and the superficial endothelium of the capillaries. This cell has been called a mesangial cell or deep cell. Because of its location and its phagocytic capabilities, it seems comparable to a pericyte or adventitial cell (Plate 47), which makes it a component of the reticuloendothelial system.

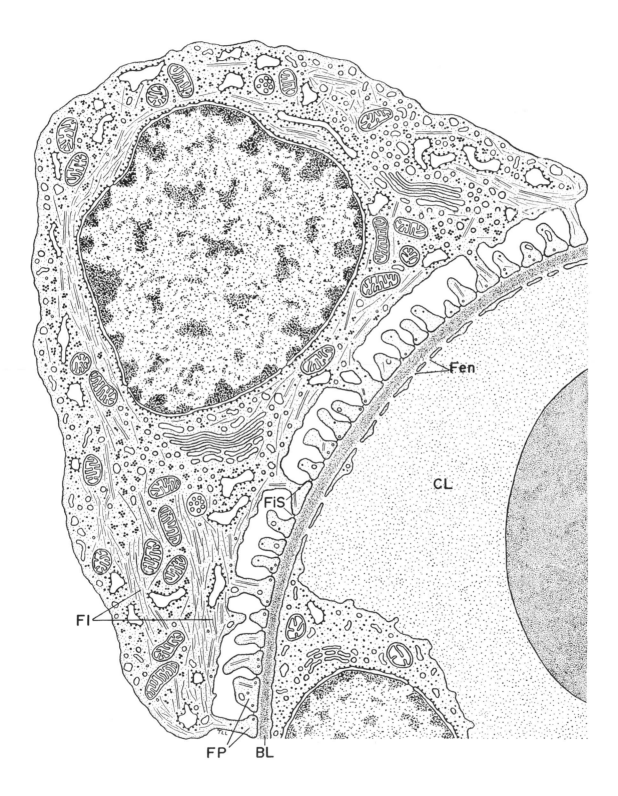

99—PROXIMAL TUBULE EPITHELIAL CELL

The visceral layer of glomerular epithelium reflects off the vessels at the vascular pole of the glomerulus and is continuous with the parietal layer, which in turn is continuous with the epithelium of the proximal tubule at the urinary pole of the nephron. The proximal tubule consists of a convoluted portion (pars convoluta) and a straight portion (pars recta), which continues into the thin limb of Henle's loop. The cells of the convoluted tubule are cuboidal. Numerous processes extend from the lateral portions of the cell, especially basally, and interdigitate with those of adjacent cells.

Many closely packed microvilli project from the apical surface and form the brush border. Apical pits (AP) and invaginations arise at the bases of the microvilli and extend into the cytoplasm. Many of these seem to pinch off, forming apical vesicles (AV) that fuse to produce vacuoles. The apical vesicles and vacuoles may give rise to some of the many lysosomes (Ly) occurring in these cells. Lysosomes and multivesicular bodies have a heterogeneous content.

The spherical nucleus lies toward the base of the cell. A Golgi apparatus is above or lateral to the nucleus. Some short cisternae of rough-surfaced endoplasmic reticulum are most numerous in the basal portion of the cell. Free ribosomes, occurring in clusters and rosettes, are found in the same area. Mitochondria are found in all regions of the cell except just beneath the surface. Basally and in the cytoplasmic processes, the mitochondria are oriented in the long axis of the cell and, in this region, are elongated with numerous cristae that extend transversely across the organelle. The matrix is dense and contains opaque granules.

Of 125 ml. of fluid filtered by the glomeruli, 124 ml. is resorbed during passage through the tubules; furthermore, the content of the remaining 1 ml. is modified. The brush border of the proximal tubule amplifies the absorptive area in contact with the filtrate in the lumen. About 85 per cent of the filtered sodium chloride and water is resorbed in this region. Sodium is removed by active transport, and the water and chloride follow passively. The mitochondria in close association with the basal processes provide a source of energy for active transport by which sodium is transported out of the cell in exchange for potassium. All of the glucose, amino acids, proteins, and other substances in the filtrate are resorbed and thus conserved. The pits and invaginations at the surface of the cell have been shown to be active in the uptake of proteins. The villi are essential in transport of glucose and may contain a carrier molecule that transfers the glucose across the membrane into the cytoplasm. Substances like urea, uric acid, and creatinine, on the other hand, are not resorbed and pass out in the urine to be eliminated from the body. The proximal tubule cells are also capable of secreting some substances, such as creatinine, into the lumen.

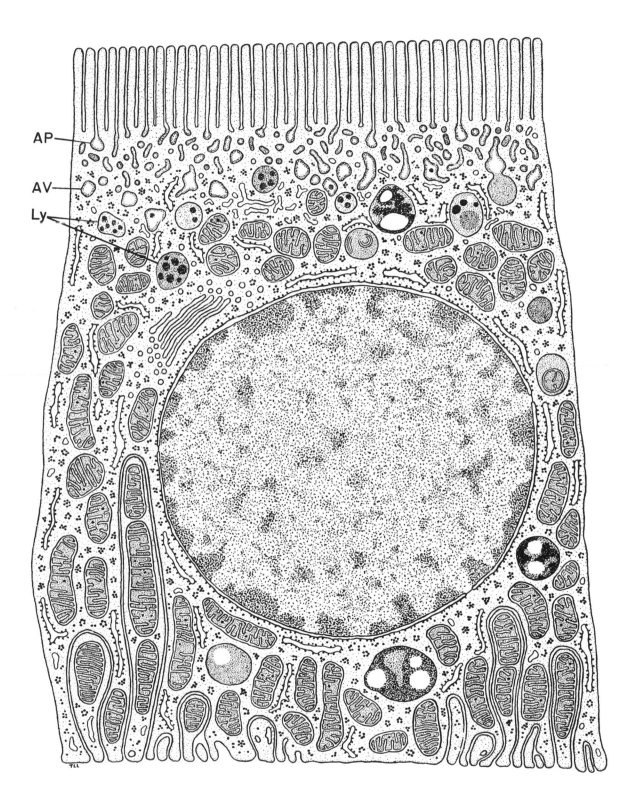

AP

AV

Ly

100—LOOP OF HENLE: THIN LIMB EPITHELIAL CELL

The loop of Henle is composed of the straight portion of the proximal tubule, the descending thin limb, the ascending thin limb, and the ascending thick limb or straight portion of the distal tubule. The thin limbs are composed of a single layer of flattened cells. The nucleus causes the central portion of the cell to bulge. The cells are highly irregular in shape with many lateral processes (top of Plate). In contrast to the basal processes of proximal tubule cells, which extend under adjacent cells, the processes of thin limb cells occupy the full thickness of the epithelium, interdigitating and alternating with one another. Junctional complexes occur between the adjacent processes.

A sparse number of microvilli, variable in length, occur on the luminal surface (right of Plate). A Golgi apparatus is present, usually lateral to the nucleus. A number of short, flat cisternae of rough-surfaced endoplasmic reticulum are found. Free ribosomes usually occur in clusters or rosettes. Mitochondria assume various forms. A small number of lysosomes and lipofuscin pigment bodies are encountered. Filaments run through the cytoplasm, and at the base of the cell they are oriented into bundles.

The loop of Henle is necessary for the production of a hypertonic urine. The wall of the descending limb is permeable to sodium and water. The ascending limb, on the other hand, is impermeable to water but extrudes sodium by active transport into the interstitial spaces of the medulla. As a result, the osmotic concentration of the interstitium is increased, and water leaves the descending limb and sodium enters by passive diffusion.

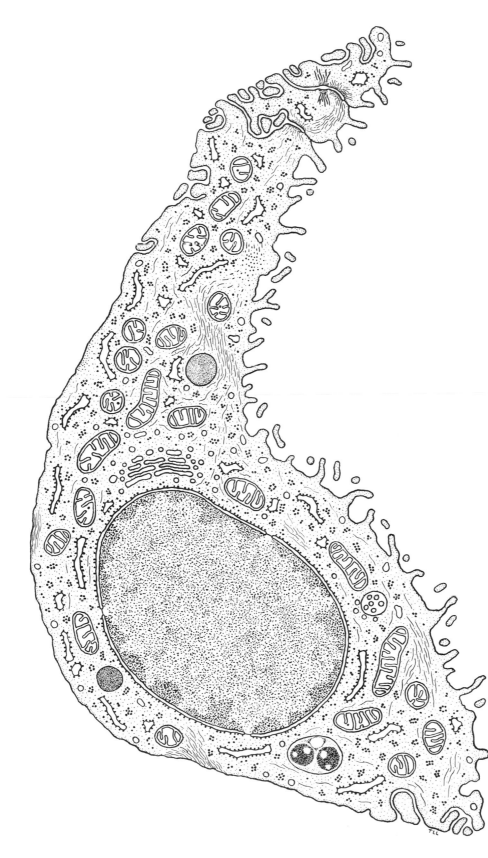

101 — DISTAL TUBULE EPITHELIAL CELL

The distal tubule arises from the thin limb of Henle's loop and is composed of a straight portion (thick ascending limb of Henle's loop) and a convoluted segment that is continuous with the collecting duct. The cells are cuboidal with the nucleus closer to the apical surface than to the basal. The nucleus contains a nucleolus and presents undulations on its luminal side. The apical surface of the cell bears a few short microvilli. The basal interdigitations of lateral processes of adjacent cells are more extensive than in the proximal tubule and often extend far up into the cell.

Large numbers of elongated mitochondria (M) fill the basal cell processes. The mitochondria are complex in internal structure, with numerous cristae. Many of the cristae have angulations along their length and extend transversely across the organelle. They often branch and anastomose, sometimes giving the appearance of a fenestrated network. The close association of the mitochondria with the large membrane surface provided by the basal processes has been interpreted as an approximation of energy source with a membrane engaged in active transport of ions. A Golgi apparatus occurs in the upper portion of the cell near the nucleus. Short cisternae of rough-surfaced endoplasmic reticulum and free ribosomes occur in the cytoplasm. A small number of lysosomes and multivesicular bodies are found. Vesicles (V) and small vacuoles are common in the apical cytoplasm. Many of the vesicles are crescent-shaped or composed of concentric membranes.

Because sodium is extruded from the ascending limb of the loop of Henle, the urine reaching the convoluted portion of the distal tubule is hypotonic, but reduced in volume. Sodium continues to be removed in the distal tubule but is replaced by other cations, such as hydrogen, potassium, and ammonia, resulting in acidification of the urine. Resorption of sodium ions is promoted by aldosterone released by the zona glomerulosa of the adrenal cortex. In the presence of antidiuretic hormone released by the posterior lobe of the pituitary, the distal tubule is permeable to water. Consequently, water leaves the tubule, further concentrating the urine.

Where the straight portion of the distal tubule comes into contact with the afferent arteriole, the cells are specialized to form the macula densa. In this region, the tubule cells have less extensive basal infoldings and fewer mitochondria, and the Golgi apparatus is situated on the side of the nucleus opposite the tubule lumen. The cells of the macula densa and the juxtaglomerular cells (Plate 103) together constitute the juxtaglomerular apparatus.

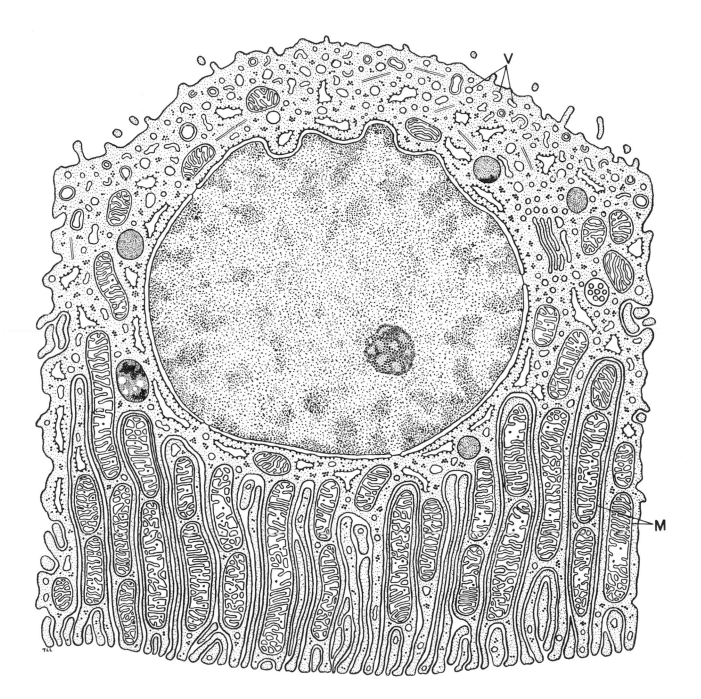

102—COLLECTING DUCT EPITHELIAL CELL

The cuboidal to columnar collecting duct cells have a central nucleus with a nucleolus. A few cytoplasmic processes interdigitate basally, and some interdigitations occur laterally. A few stubby microvilli project into the lumen. Organelles are, for the most part, sparse in number. The endoplasmic reticulum is represented by a few short, ribosome-studded cisternae. A small Golgi apparatus is located adjacent to the nucleus. Vesicles are distributed throughout the cell but are most numerous in the apical portion. Round to oval mitochondria are scattered randomly in the cell. A small number of lysosomes and multivesicular bodies are found. Clusters of ribosomes are scattered through the cytoplasm. Microtubules often run parallel to the plasma membrane. Filaments also occur, terminating in bundles on the basal membrane. Most collecting duct cells contain the general features just described. There are, however, differences in cell structure in different regions of the duct. Progressing from the initial part of the collecting duct in the cortex to the medulla of the kidney, the cells change in shape from low-cuboidal to columnar. The cell borders become less elaborate, and organelles become less numerous.

A second cell type is found in the first portion of the duct. This cell, called a dark cell, has a denser hyaloplasm than does the more frequently occurring light cell, described above. Organelles and basilar interdigitations are more numerous in the dark cell, but in most respects they are similar to those in the light cell.

Urine entering the collecting duct is still hypotonic or isotonic with respect to the plasma. Like the distal tubule, the collecting duct is permeable to water in the presence of antidiuretic hormone. Water passively diffuses from the tubule, leaving a concentrated urine.

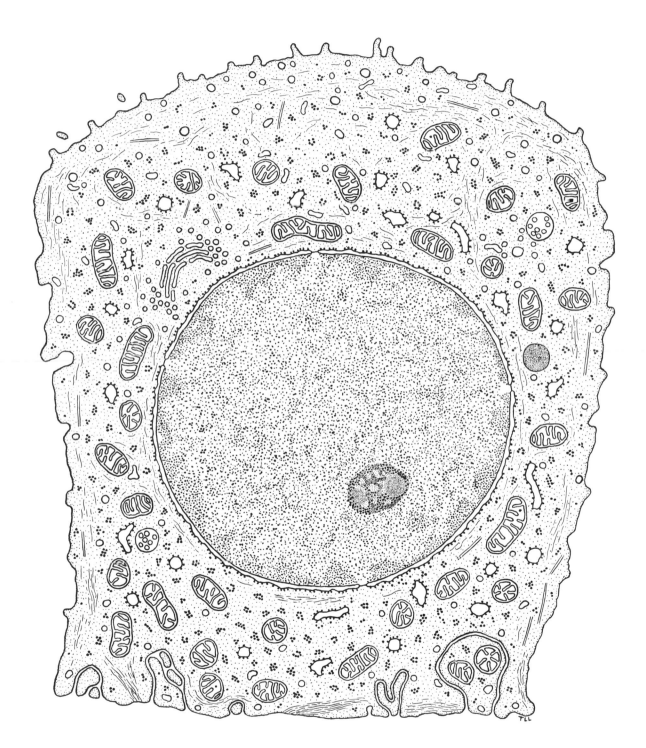

103 — JUXTAGLOMERULAR CELL

Juxtaglomerular cells are found along the course of the afferent arteriole just before its entrance into the glomerulus. These cells replace the smooth muscle cells in the wall of the arteriole and border the endothelium. They are highly irregular in shape, and they contain conspicuous secretory granules (SG). The most common type of granule is large, spherical, and composed of moderately dense, homogeneous material. Some of the spherical granules appear to coalesce, forming irregular aggregates. The second granule is oval or rod-shaped and has a crystalline (Cry) internal structure. These secretory granules are thought to contain the vasopressor substance renin, a proteolytic enzyme.

The granules originate in a well-developed Golgi apparatus (G) located near the nucleus. Some of the larger Golgi vacuoles contain material similar to that comprising the large granules. The cells also have an extensive system of rough-surfaced endoplasmic reticulum. Many of the cisternae are dilated and saccular. Oval mitochondria are distributed through the cytoplasm.

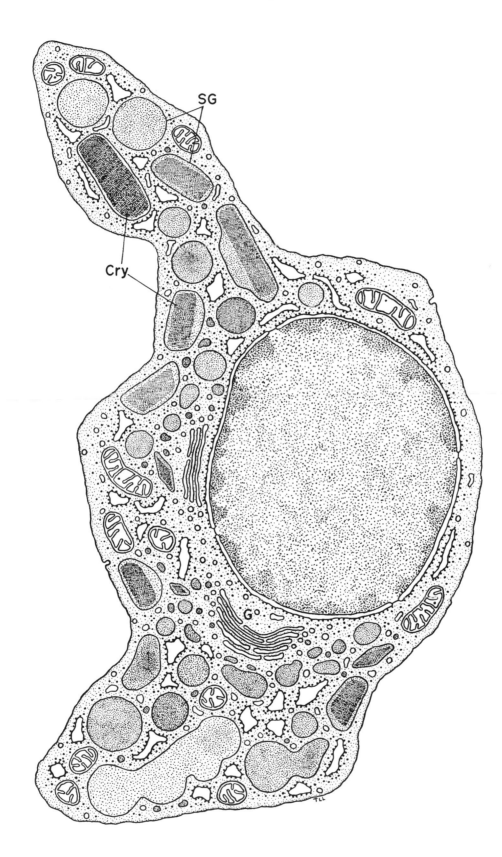

URINARY SYSTEM

104—TRANSITIONAL EPITHELIAL CELL

The calyces and pelvis of the kidney, ureter, and bladder are lined by transitional epithelium. This epithelium is capable of undergoing rapid changes in surface area, depending on the degree of distention of the organ. The superficial squamous cells, illustrated here, are polyhedral. They become columnar when the viscus is collapsed and flattened when it is distended. The surface contour (left of Plate) is convex and has many crests and angulations. The surface ectoplasm is filled with bundles of filaments (Fl) but otherwise contains few organelles. The filaments generally course parallel to the surface of the cell and insert on junctional complexes that occur on the lateral borders. The lateral and basal membranes have numerous interlocking interdigitations with adjacent cells. The cells have an extensive Golgi system near the nucleus. Vesicles are common throughout the cytoplasm. Other cytoplasmic organelles are mitochondria, lysosomes, and multivesicular bodies.

A characteristic structure found throughout the cell is a fusiform vesicle (FV) that usually assumes a crescent-shaped or cigar-shaped membranous profile. Other vacuoles are more dilated, oval, or spherical. One function that has been suggested for the fusiform vesicle is storage of surface membrane, so that it can be mobilized quickly when the bladder distends. The size of the clefts on the free surface indicates that they could give rise to the fusiform vesicles when compressed. Alternatively, the fusiform vesicles might represent stages in the eventual destruction of surface membrane. Constant replacement of membrane might be related to the membrane's impermeability to water. This impermeability is necessary to prevent water from being drawn out of the cells by the hypertonic urine.

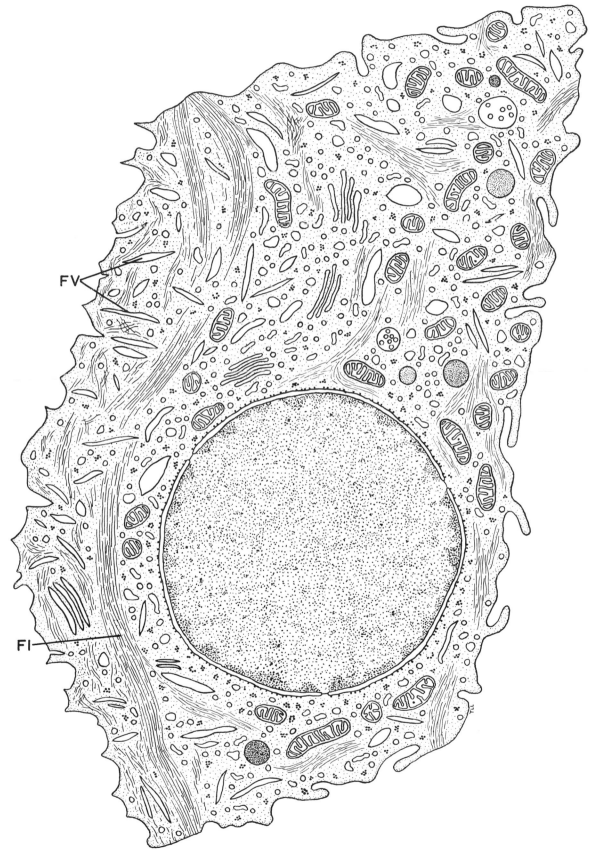

REFERENCES

Barajas, L. and H. Latta, 1963. A three-dimensional study of the juxtaglomerular apparatus in the rat. Lab. Invest., 12:257–269.

Bulger, R. E., C. C. Fisher, C. H. Myers, and B. F. Trump, 1967. Human renal ultrastructure. II. The thin limb of Henle's loop and the interstitium in healthy individuals. Lab. Invest., 16:124-141.

Fisher, C. C., R. E. Bulger, and B. F. Trump, 1966. Human renal ultrastructure. I. Proximal tubule of healthy individuals. Lab. Invest., 15:1357-1394.

Hicks, R. M., 1965. The fine structure of the transitional epithelium of rat ureter. J. Cell Biol., 26:25-48.

Jorgensen, F., 1967. Electron microscopic studies of normal visceral epithelial cells. Lab. Invest., 17:225-242.

Lee, J. C., S. Hurley, and J. Hopper, 1966. Secretory activity of the juxtaglomerular granular cells of the mouse. Lab. Invest., 15:1459-1476.

Myers, C. E., R. E. Bulger, C. C. Fisher, and B. F. Trump, 1966. Human renal ultrastructure. IV. Collecting duct of healthy individuals. Lab. Invest., 15:1921-1950.

Osvaldo, L. and H. Latta, 1966. The thin limbs of the loop of Henle. J. Ultrastruct. Res., 15:144-168.

Rosen, S. and C. C. Fisher, 1968. Observations on the rhesus monkey glomerulus and juxtaglomerular apparatus. Lab. Invest., 18:240-248.

MALE REPRODUCTIVE SYSTEM

MALE REPRODUCTIVE SYSTEM

105 — TESTIS: SPERMATOCYTE

The cytoplasmic structure of the spermatocyte, as well as the spermatogonium, is unspecialized like that of undifferentiated cells. Membranous elements are sparse and consist of short, scattered rough-surfaced cisternae, a small Golgi apparatus, and a few isolated smooth membranes or vesicles. Free ribosomes, on the other hand, are more common and often occur in clusters or rosettes. Mitochondria (M) are small, and oval or round in profile. The cristae tend to be dilated with a clear intracristal space. The mitochondrial matrix, however, is relatively dense. A pair of centrioles (Ce) is found near the nucleus and Golgi apparatus.

Division of spermatogonia (giving rise to spermatocytes) and of spermatocytes is incomplete, so that the daughter cells remain connected by intercellular bridges (IB). These are broad continuities of cytoplasm between adjacent cells. The intercellular space is slightly widened on each side of the bridge, and the membrane encircling the bridge is thickened. It is apparent that transfer of substances can readily occur across these areas of continuity, and it is thought that the bridges are responsible for the precise synchrony of developmental events occurring in joined cells during spermatogenesis.

The nucleus of the spermatocyte shows some characteristic features during meiosis. One large nucleolus (Nl) and several smaller nucleoli are present during the long prophase of the first maturation division. The main nucleolus is composed of a large round body, with an outer region of dense granules and a dense fibrillar core. This portion of the nucleolus is connected to a dense chromatin mass by a stalk of lower density. The chromatin is closely applied to the inside of the nuclear envelope. Synaptonemal complexes (SnC) formed by the pairing and fusion of homologous chromosomes are located within the chromatin mass and are connected to the nuclear envelope. These structures consist of a pair of coarse filaments separated by a clear space that is bisected by a dense line. On the opposite side of the nucleus, the sex vesicle (SxV) is found. This consists of a mass of condensed chromatin containing two filaments, one long and one short, representing axial structures of the X and Y chromosomes. Accumulations of dense material are associated with the filaments along their lengths. The two filaments approach one another to form a synaptonemal complex that terminates on the nuclear envelope.

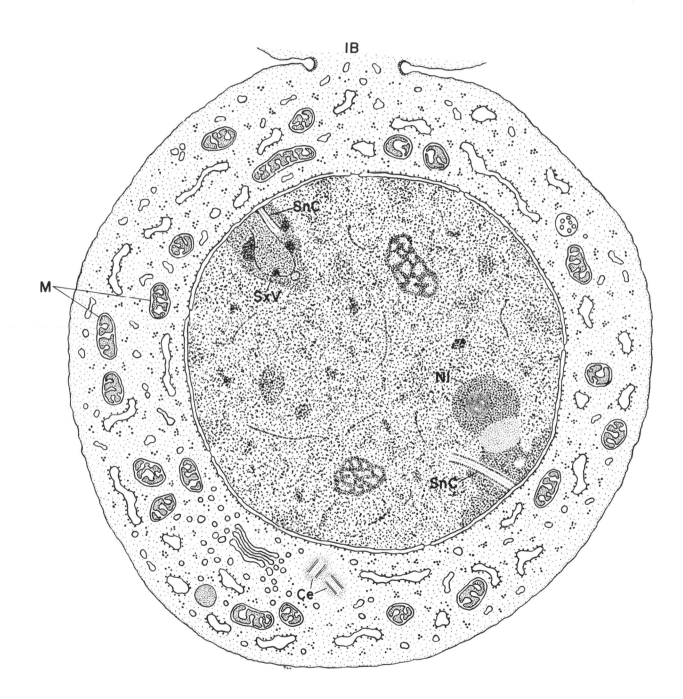

MALE REPRODUCTIVE SYSTEM

106—TESTIS: SPERMATID

Spermiogenesis is the process of differentiation of the spermatid into the mature sperm. Immature spermatids are uncomplicated in structure; the cytoplasm is similar to that of spermatocytes, but with a few differences. The intracristal space of the mitochondria (M) shows further dilation. Ribosomes are not as densely packed in the cytoplasm. A few lysosomes and multivesicular bodies are seen at this time.

The cells soon undergo a series of changes related to the formation of the acrosome and the tail. At the anterior pole of the cell, the Golgi apparatus becomes enlarged, and associated with it are membrane-bounded granules. These granules are rich in carbohydrates and are known as proacrosomal granules (PaG). The individual granules coalesce to form a large dense acrosomal granule (AcG). The acrosomal granule is situated within a dilated acrosomal vesicle (AcV) containing particulate material. The acrosomal vesicle becomes closely applied to the apical pole of the nucleus, and the space between the inner and outer nuclear membranes of the envelope is reduced to a thin line adjacent to the area of contact. Small Golgi vesicles continue to fuse with the acrosomal vesicle, causing it to enlarge. Later, the acrosomal vesicle and its contained acrosomal substance spread until they cover the anterior two thirds of the nucleus (see Plate 108).

During spermiogenesis, the nucleus elongates and the chromatin becomes condensed. At this stage, many fine filaments are seen in the nucleoplasm. These aggregate into coarser fibrils that shorten and condense into coarse, irregular dense granules. Later, the dense granules become compacted into a dense homogeneous mass.

In the early spermatid, the two centrioles migrate from the Golgi zone to the caudal end of the cell and become positioned near the cell surface. The distal centriole, which later disappears, is oriented perpendicular to the cell surface and gives rise to the flagellum, which has the usual nine-plus-two organization of fibrils. The proximal centriole, which persists in the mature spermatozoon, is at right angles to the distal centriole and is closer to the nucleus. Overlying the proximal centriole is a thin lamina of dense material that is the anlage of the articular surface of the future connecting piece. The appearance of the striated columns (StC) of the connecting piece is marked by a row of rectangular densities alongside the centrioles. A large spheroidal mass of finely granular material, or paracentriolar body (PcB), is found on one side of the distal centriole. The centrioles and base of the flagellum sink inward to associate with an indentation in the caudal end of the nucleus. Dense material representing the early basal plate (BP) is applied to the outer membrane of the nuclear envelope. A small mass of dense material is also applied to the inner aspect of the cell membrane where it is reflected onto the flagellum at the base of the indentation. This mass, called the ring centriole (RC), persists as the annulus between the middle piece and principal piece of the mature sperm tail. The accumulation of less dense substance with the annulus is the chromatoid body (CB). Finally, in the caudal cytoplasm are some longitudinally arranged microtubules (Mt) that later form the manchette. The manchette is a group of microtubules extending from a C-shaped configuration of the plasma membrane just behind the caudal end of the acrosomal cap to the posterior cytoplasm where the microtubules encircle the flagellum.

As the nucleus elongates during spermiogenesis, the cytoplasm becomes displaced caudally. The postnuclear cytoplasmic lobe contains cytoplasmic organelles including the Golgi apparatus and multivesicular bodies rich in acid phosphatase. Just before the spermatids are released as spermatozoa, the cytoplasmic mass is cast off as a residual body which is degraded.

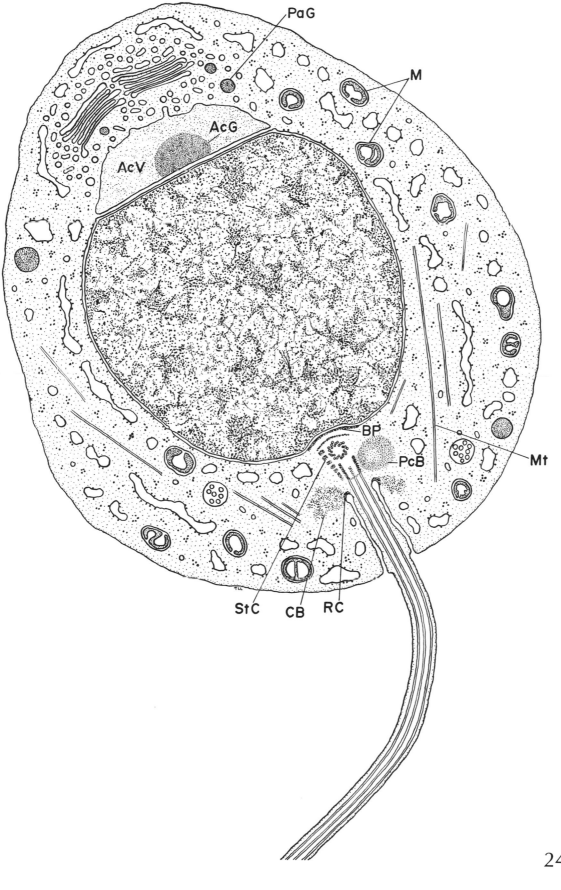

MALE REPRODUCTIVE SYSTEM

107—TESTIS: SPERMATOZOON

The spermatozoon consists of two principal regions, the head containing the nucleus and the motile tail. The tail can be further subdivided into four regions, the neck (Nk), middle piece (MP), principal piece (PP), and end piece (EP). In the illustration, the middle piece is longer and the principal piece considerably shorter in relation to the other regions. The shape of the head differs in different species and is largely determined by the shape of the nucleus. The head is usually flattened and here is shown from the side, but the neck is viewed face on. The nucleus (N) occupies most of the head and is extremely dense and homogeneous except for occasional clear cavities or nuclear vacuoles. The condensed chromatin is closely bounded by the nuclear envelope. The anterior portion of the nucleus is enclosed by the acrosomal cap or acrosome (Ac). It is composed of moderately dense material and is thickest at the tip of the head but extends back in a thinner layer over the anterior two thirds of the nucleus. The acrosomal material is completely enclosed by a membrane closely applied to the outer membrane of the nuclear envelope except at the tip. Here, the membrane is infolded, leaving a larger space, the subacrosomal space, between it and the nuclear envelope. Caudal to the posterior margin of the acrosomal cap, the plasma membrane is separated from the nuclear envelope by a thin dense layer.

The neck (Nk) of the spermatozoon is the region between the head and the first gyre of the mitochondrial helix of the middle piece. Its principal component is the connecting piece, which attaches the tail to the head. The connecting piece consists of nine striated columns (StC) that fuse along an oblique line with the outer dense fibers (ODF) of the middle piece. The columns have a cross-banded or segmented appearance. Anterior to the junction of the columns with the fibers, two of the striated columns on each side of the connecting piece fuse to form two major columns, leaving five smaller minor columns. One of the major columns is expanded at its anterior end to form a head or capitellum (Cap), which is fused with the basal plate (BP). The other major column turns as it approaches the head and merges with the capitellum. Together, the major columns form at the base of the head an articular structure upon which the minor columns converge. The articular structure rests in a shallow concavity or implantation fossa lined by the nuclear envelope at the caudal end of the nucleus. A layer of dense, amorphous material comprising the basal plate (BP) is applied to the outer surface of the nuclear envelope. Lateral to the basal plate, the nuclear envelope (NE) is reflected away from the condensed chromatin and extends back as a fold into the neck region. A centriole (Ce) is located in the interior of the connecting piece just below the articular surface. The centriole is set obliquely within a niche in one of the major columns. Finally, two or more irregular, longitudinally oriented mitochondria are found in the neck region.

The middle piece (MP) of the tail is about one and a half times as long as the head and has a fusiform enlargement, or cytoplasmic droplet, midway along its length. The middle piece contains elongated mitochondria (M) wound helically around the flagellar fibers (F). The mitochondria are situated end to end in the helix. They are round in cross section with lamellar cristae extending into a dense matrix. The cytoplasmic droplet contains a number of smooth vesicles, vacuoles, and flattened saccules. Caudal to the last gyre of the mitochondrial sheath is the annulus (An). This structure is a dense ring to which the flagellar membrane is closely applied and which marks the end of the middle piece.

The axial filament complex of the sperm flagellum is composed of two central single fibrils surrounded by nine double fibrils (see cross section, upper left of Plate). The central fibrils are circular with a less dense interior. The two subfibrils of the outer doublets have been designated A and B. Subfibril A is circular in profile, while the wall of subfibril B is not complete but C-shaped. Subfibril A is slightly smaller, it lies closer

(Continued)

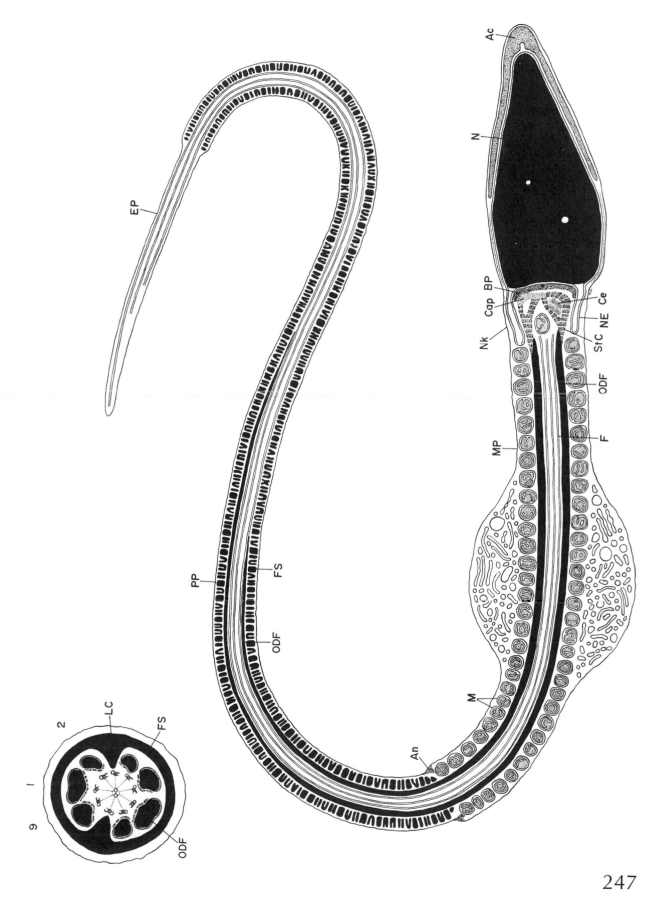

107—TESTIS: SPERMATOZOON (*Continued*)

to the center of the flagellum, and its center is denser. Two short divergent arms, of which the outermost is more prominent, project from the surface of subfibril A. The arms contain a protein, dynein, which has ATPase activity and which could provide energy for ciliary movement by means of hydrolysis of ATP. The two central fibrils are connected by a thin, short, straight line and two arcs on the outside. From the arcs on the central fibers, radial lines extend to subfibril A of each doublet. The fibrils of the axial filament seem to end blindly in the interior of the connecting piece.

The axial filament is surrounded by a group of nine, outer dense fibers (ODF). These fibers are thick and shaped like flower petals with the broad end facing outward. They are extremely dense except for a outer cortical rim of lower density. Dense profiles or satellite fibrils are aligned along the inner and lateral aspects of the fibers in the middle piece. Usually, two of the fibers on one side and one on the opposite side are larger than the others. The single large fiber has been designated as number 1, and the others are numbered consecutively in the direction of the arms on the subfibrils A of the axial filament complex. The fibers taper, progressing caudally, and terminate at different points along the tail. Their terminals are fused with the corresponding doublet of the axial complex. In the cross section, two of the outer fibers, numbers 3 and 8, have terminated, leaving seven fibers.

The long principal piece (PP) of the tail is characterized by a fibrous sheath (FS) surrounding the axial fibrils. The fibrous sheath is composed of circumferentially arranged fibers or ribs. On opposite sides of the tail the ribs fuse with two longitudinal columns (LC) which occupy the regions previously occupied by the two outer fibers that have terminated. Progressing caudally, the fibrous sheath becomes thinner until it terminates abruptly, marking the end of the principal piece.

The end piece (EP) is similar in structure to a cilium with only the limiting membrane enclosing the axial filament complex. Near the tip, the doublets are reduced to single fibers that end at different levels.

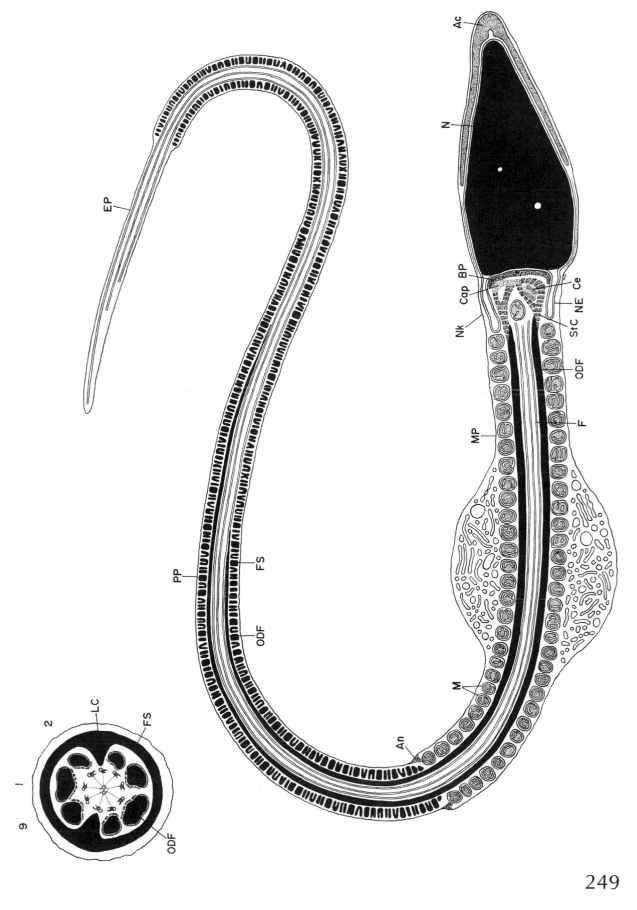

249

MALE REPRODUCTIVE SYSTEM

108—TESTIS: SERTOLI CELL

The Sertoli cells are crowded between the germ cells of the seminiferous tubules and apparently perform supporting and nutritive functions. The cells rest on the basement lamina and extend to the surface of the tubule. They are highly irregular in shape, being indented by neighboring cells and containing within deep recesses the heads of maturing spermatozoa. The nucleus lies in the middle or basal portion of the cell and is roughly oval in shape but has deep indentations of the surface. The nucleus contains a principal nucleolus (Nl) flanked by two accessory nucleoli (AN). The principal nucleolus consists of a large nucleolonema and several smaller, round amorphous parts. The accessory nucleoli are smaller, round bodies composed of highly condensed material.

The general hyaloplasm is of low density and has a number of filaments coursing through it. The rough-surfaced endoplasmic reticulum is not prominent, but there are a number of smooth-surfaced tubules, channels, and vesicles comprising the smooth endoplasmic reticulum. A few ribosomes are dispersed in the cytoplasm. An annulate lamella (AL) is often seen near the nucleus. This structure consists of smooth, fenestrated cisternae arrayed concentrically around a lipid droplet or pigment granule. At its periphery, the lamellae are continuous with ribosome-studded cisternae. The Golgi apparatus is relatively small and occurs in association with a pair of centrioles near the nucleus. A cluster of tightly packed vesicles (V) is sometimes seen as well. The mitochondria are small rods with a small number of transverse cristae and a dense matrix. Lysosomes, lipid droplets, and lipofuscin pigment granules are fairly common. The latter are often large and consist of a limiting membrane enclosing a large lipid droplet and a cap of cytoplasm with smaller droplets and dense material. A large, needle-shaped crystalloid (Cry) is found near the level of the nucleus. The crystalloid consists of a parallel array of regularly spaced coarse filaments.

Some unusual specializations are found at the surface of Sertoli cells. Where Sertoli cells are adjacent to spermatogonia, spermatocytes, or early spermatids, the usual 150 to 200 Å intercellular space is found (lower right of Plate). However, where two Sertoli cells adjoin, the intercellular space is reduced to 70 to 90 Å. Just inside the membrane are periodic densities formed by aggregations of fine filaments (Fl). Deep to the band of filaments are flat cisternae of endoplasmic reticulum (SsC) running parallel to the membrane. The inner or cytoplasmic surface of the cisternae bears ribosomes, while the outer surface facing the filamentous layer is smooth. Similar specializations are found in the Sertoli cell cytoplasm adjacent to where mature or nearly mature sperm heads lie in the deep recesses. In this region, though, the subsurface cisternae have no associated ribosomes. The surface specializations of Sertoli cells have some similarities with *zonulae adherentes* or *maculae adherentes*, especially in the association of dense material with the membranes. Thus, these junctions could play a role in adhesion and maintenance of the structural integrity of the epithelium. The Sertoli cell–sperm junction would constitute a half or hemidesmosome. Furthermore, the narrow intercellular gap is a type of close junction over which cell-to-cell communication can occur. It has been suggested that these connections could play a role in the synchronization of regional development characteristic of the spermatogenic wave (Flickinger and Fawcett).

A late spermatid is embedded in the apical cytoplasm of the Sertoli cell. The nucleus (N) has a flattened pyriform shape and the nucleoplasm has condensed into a dense homogeneous mass. The acrosome (Ac) is closely applied to the nucleus and covers its anterior portion.

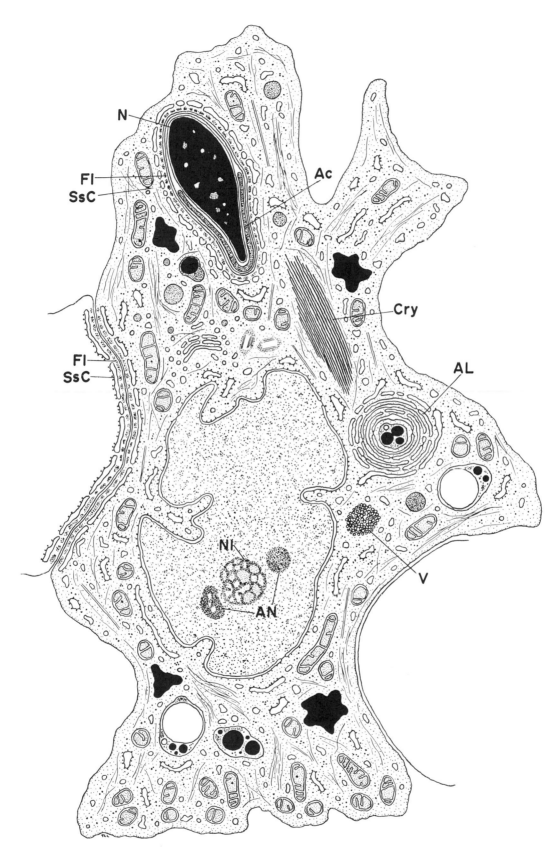

109—TESTIS: INTERSTITIAL CELL

The interstitial cells or cells of Leydig are polyhedral cells lying singly or in small groups between the seminiferous tubules of the testis. The nucleus is oval in shape and eccentrically located. The most conspicuous feature of the cytoplasm is the abundant smooth-surfaced endoplasmic reticulum (SER). In the central regions of the cell, the agranular endoplasmic reticulum is in the form of branching, irregular, and anastomosing tubules. Toward the periphery, the reticulum more often is in the form of closely packed, fenestrated cisternae. Sometimes the cisternae are arranged in concentric whorls around a lipid droplet (LD). Here and there are a few small stacks of rough-surfaced endoplasmic reticulum (ER) connected with the agranular reticulum. Free ribosomes are relatively abundant in the interstices of the reticular elements, and some glycogen granules are also present.

Mitochondria (M) are abundant and are generally oval to elongated in shape. The cristae are lamellar to tubular in shape and extend into a dense matrix. Small Golgi complexes are dispersed in the cytoplasm. Large lipid droplets, lipofuscin pigment granules, and a few microtubules occur. In the human, the interstitial cells contain crystals (of Reinke) apparently composed of closely packed microtubules (not shown).

Interstitial cells are the main source of testicular androgenic steroid hormones, and their structure should be compared with other steroid-secreting cells, especially lutein cells (Plate 119) and adrenocortical cells (Plates 146 to 148). The membranes of the smooth-surfaced endoplasmic reticulum are thought to contain the enzymes necessary for the synthesis of androgens from pregnenolone, including 17-hydroxylase and 17-desmolase. Pregnenolone is formed in the mitochondria from cholesterol.

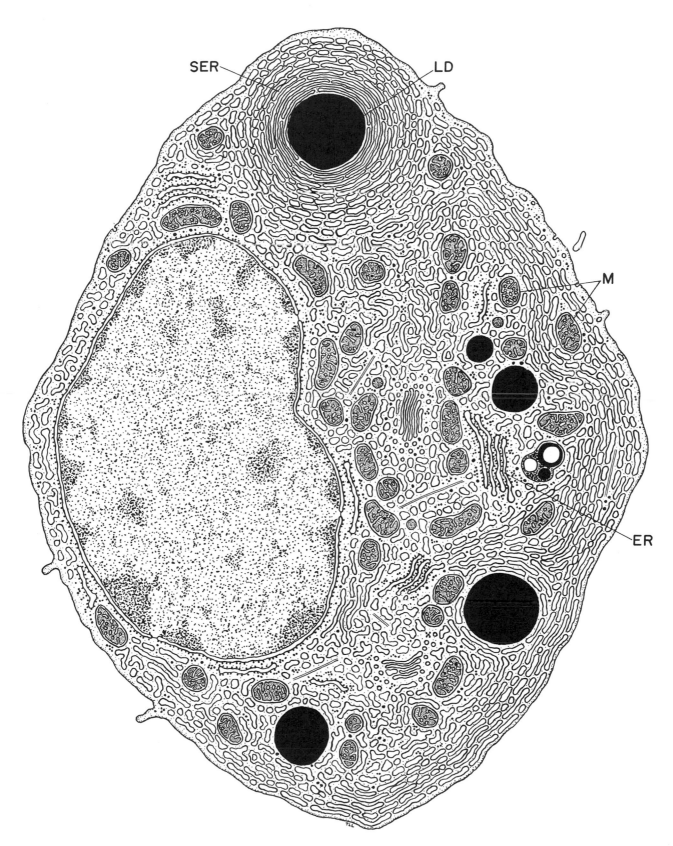

110—DUCTULI EFFERENTES: CILIATED CELL

The epithelium of the ductuli efferentes is composed of two intermingling cell types: a ciliated and a non-ciliated cell. The cells vary in height from cuboidal to columnar, giving the lumen of the ducts a festooned outline.

The ciliated cells have a basal nucleus and possess a tuft of cilia on the apical surface. A few microvilli are interspersed between the cilia. Near the apex of the cell, small vesicles are abundant. The cells are relatively uncomplicated in cytoplasmic structure. They have a small Golgi complex above the nucleus. Elements of endoplasmic reticulum are sparse, but free cytoplasmic ribosomes are common. Mitochondria are also numerous and are round or rod-shaped in profile. The cells also contain dense bodies or lysosomes as well as a number of lipofuscin pigment granules. The lateral borders and the basal surface, which rests on a basement membrane, are straight.

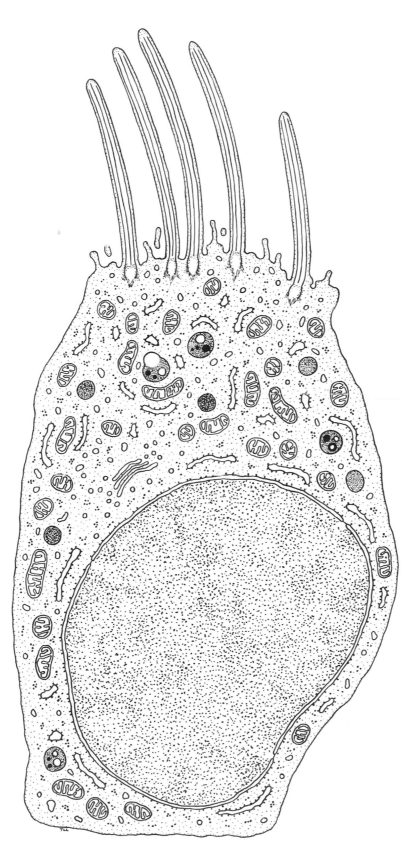

111—DUCTULI EFFERENTES: NON-CILIATED CELL

The non-ciliated cell of the efferent ductule is more complicated than the ciliated cell. Like the latter, the non-ciliated cell is columnar. Its nucleus, which is indented and contains a nucleolus, is situated in the base of the cell. The apical surface is rounded and from it project a number of microvilli. The surface plasma membrane bears a surface extraneous coat of filamentous material. Small vesicles or apical pits (AP) are in continuity with the plasma membrane at the bases of microvilli. These vesicles, along with the extraneous coat, appear to pinch off from the surface to lie in the apical cytoplasm as apical vesicles (AV). Also abundant in this region are elongated channels (Ch) containing moderately dense material probably derived from the extraneous coating. Larger vacuoles (Vac) containing particulate material are also seen in the upper portion of the cell. That these vacuoles are formed by fusion of the small vesicles is suggested by images in which the small vesicles are continuous with the vacuole.

A well-developed Golgi complex occurs in a supranuclear position. Elongated cisternae of rough-surfaced endoplasmic reticulum are seen, as are numerous free ribosomes. Mitochondria are abundant and similar to those in ciliated cells. Dense bodies, most of them resembling lysosomes, occur in large numbers. Finally, a few filaments run in the cytoplasm between organelles.

The surface coating of the non-ciliated cells and the abundance of vesicles is characteristic of absorbing cells, and it is thought that these cells are active in uptake of substances, particularly fluids. The apical pits and invaginations lined by a layer of filamentous material or having a bristle coat on their cytoplasmic surface are usually found in cells taking up proteins (see Plates 12, 19, 99, 112, 131). A secretory function has also been suggested for the cells, although most of the cytoplasmic granules appear to be lysosomes.

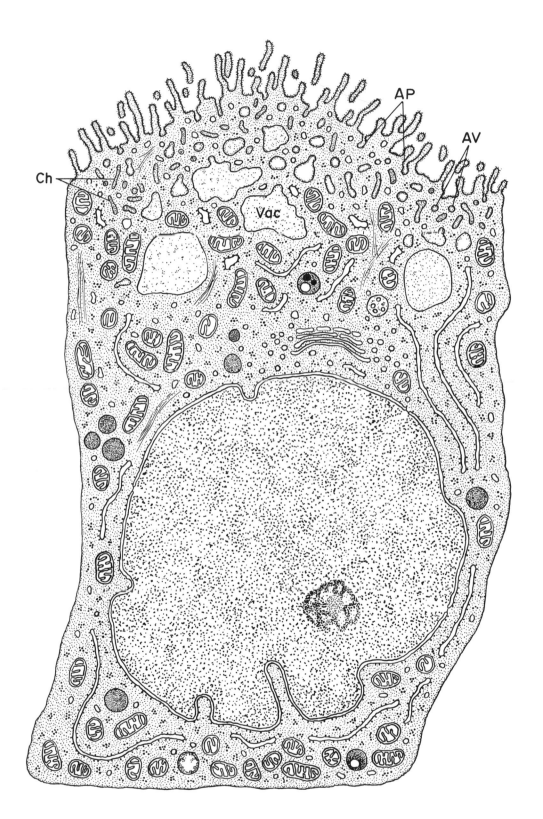

MALE REPRODUCTIVE SYSTEM

112—DUCTUS EPIDIDYMIDIS EPITHELIAL CELL

The epithelium lining the epididymis is composed of tall columnar cells and a row of small basal cells (see Plate 114). The epithelial cells of the epididymis and the ductus deferens are characterized by the presence of stereocilia (Sc) on the luminal border. The stereocilia are very long microvilli that extend three to four times the length of the usual brush border. They lack the axial filaments of cilia and are non-motile. Small pits and invaginations occur at the bases of the villi and coated vesicles are found in the apical cytoplasm. A major function of the epididymis is probably reabsorption of fluids and proteins.

Another distinguishing feature of epididymal epithelial cells is the occurrence of an extremely large Golgi complex (G) in the apical cytoplasm. This structure consists of several arrays of closely packed lamellae and a number of small vesicles. Found in the same region are large membrane-bounded granules. The content of these structures is dense and finely granular in texture and may include opaque droplets. These bodies were once considered secretory granules but it seems more likely that they represent lysosomes and lipofuscin pigment granules. Multivesicular bodies, another type of lysosome, are found in the same region.

The cells also have an extensive smooth-surfaced endoplasmic reticulum (SER), which appears as vesicular profiles filling most of the cytoplasm. Rough-surfaced cisternae are few in number and occur most frequently near the nucleus. A few ribosomal clusters are interposed among other structures in the cytoplasm. Mitochondria are usually rod-shaped with transverse, lamellar cristae. The nucleus is located toward the bottom of the cell and contains a nucleolus. There are masses of clumped heterochromatin, especially against the inside of the nuclear envelope. In addition, the nucleus of some species contains some unusual inclusion bodies (NI). They are spherical and composed of dense material; they may be free in the nucleoplasm or enclosed in a membrane. Smaller bodies may represent stages in the formation of larger globules.

The epididymal secretion contains a number of enzymes, including glycosidases, and has a high content of glycerylphosphorylcholine, but it lacks glycolysable sugars which induce sperm motility. Immobile sperm are stored in the epididymis where they undergo final maturation.

DUCTUS EPIDIDYMIDIS EPITHELIAL CELL

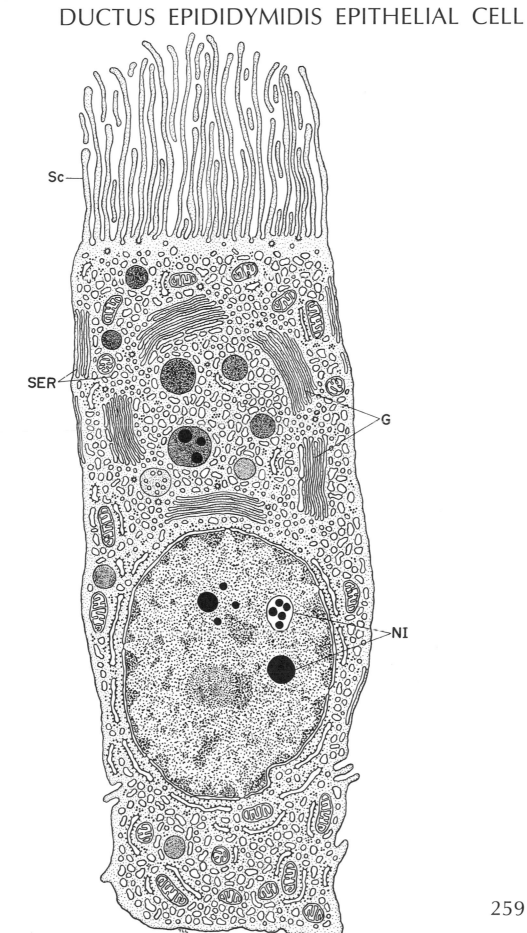

Sc

SER

G

NI

259

113 — SEMINAL VESICLE: SECRETORY CELL

The epithelium of the seminal vesicle is composed primarily of high columnar cells. The apical portions of the cell bulge far into the lumen. The free surface is provided with irregularly spaced microvilli of different lengths. An intercellular canaliculus (Ca) that extends down to the beginning of the junctional complex separates apical portions of adjacent cells. The lateral and basal borders of the cells show interdigitations.

Nuclei occur at different levels from the middle to the base of the cells. The nucleus contains a nucleolus and masses of condensed chromatin. The rough-surfaced endoplasmic reticulum is extensive, especially toward the base of the cell. Cisternae are flattened, elongated, and arranged in parallel. In the vicinity of the Golgi apparatus the cisternae are short but dilated and contain flocculent material. Ribosomes also occur free in the cytoplasm. The Golgi complex occupies a large region of cytoplasm above the nucleus. Besides flattened cisternae and vesicles, small vacuoles occur in association with this organelle. The vacuoles contain flocculent or fine particulate material within which is a core of dense material, and toward the apex of the cell are much larger secretory vacuoles (Vac) containing a large dense granule lying within the flocculent material. Squeezed between the endoplasmic reticulum are mitochondria and a few lysosomes.

The seminal vesicles secrete protein into the seminal fluid and the fine structure of the cells is comparable to that of other protein-secreting cells. The clear halo, however, around the dense granule in the secretory vacuoles is somewhat unusual. Other products, including fructose, sorbitol, citric acid, and prostaglandin, are produced by these cells and some of these could reside in the perigranular space. The seminal vesicle secretion is less acidic than the secretion of the prostate and it may be alkaline. It is pigmented due to the presence of flavins and has a high content of reducing substances.

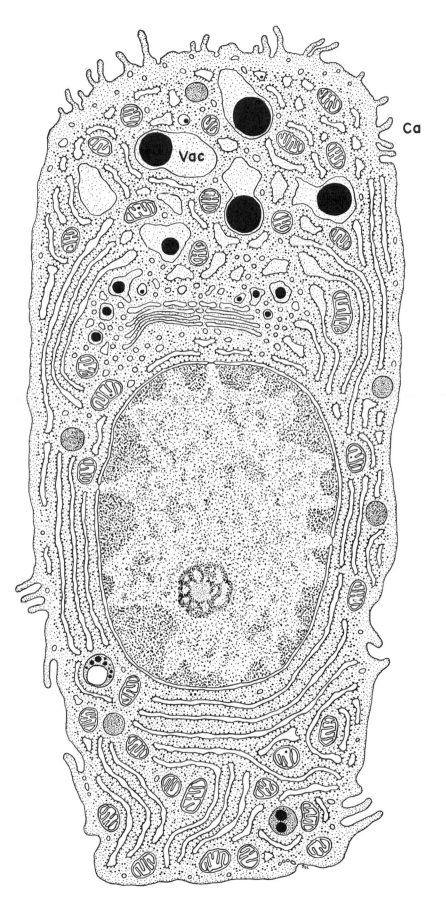

Vac

Ca

114—SEMINAL VESICLE: BASAL CELL

Basal cells are found in the epithelium between the bases of the columnar secretory cells. They occur in all the pseudostratified epithelia of the excretory ducts of the male reproductive system. They are small, stellate cells with a central nucleus. The nuclear chromatin is highly condensed and a nucleolus is lacking. The cytoplasm is relatively unspecialized. The hyaloplasm itself is of low density. A small Golgi apparatus occurs near the nucleus. A few ribosomes and vesicles are seen, but the endoplasmic reticulum is sparse. There are some mitochondria, lysosomes, and occasionally a lipid droplet. A few filaments are distributed in the cytoplasm. Basal cells in general have been regarded as undifferentiated reserve cells capable of giving rise to principal cells. However, during development they begin to appear after principal cells and thus could alternatively represent dedifferentiating or expended secretory cells. Whatever their nature, they are similar to basal cells found in other epithelia.

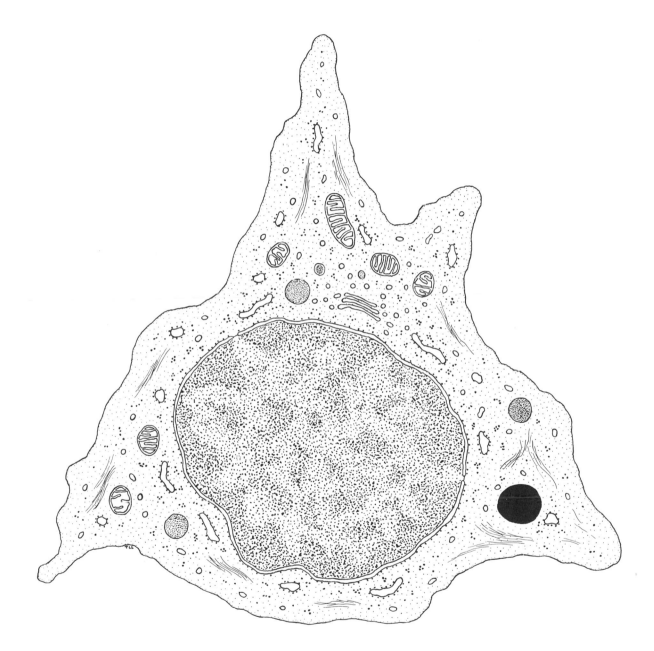

115—PROSTATIC EPITHELIAL CELL

The prostate is lined by tall columnar cells whose structure is typical of that of secretory cells. Much of the cytoplasm is filled with elongated cisternae of rough-surfaced endoplasmic reticulum. The cisternae are slightly dilated and contain flocculent material. Many ribosomes also occur free in the hyaloplasm. An extensive Golgi zone is seen above the nucleus. It consists of stacks of smooth lamellae and clusters of small vesicles. Larger, irregularly shaped vacuoles containing moderately dense material represent early secretory granules. Mature secretory granules (SG) are round and more common toward the apex of the cell. The prostate produces a thin milky fluid that is slightly acidic (pH 6.5) and contains a number of enzymes, including diastase, β-glucuronidase, and proteolytic enzymes such as fibrinolysin and acid phosphatase. The fluid also contains citric acid, which is beneficial to sperm motility.

The free surface of the prostatic cells bears short, stubby microvilli. The apical rim of cytoplasm is largely devoid of organelles. Lateral borders are relatively straight with a few interdigitations toward the base. The nucleus is found in the base of the cell and contains a nucleolus. Mitochondria are generally elongated and not unusual in structure. A few lysosomes and lipid droplets are also seen.

SG

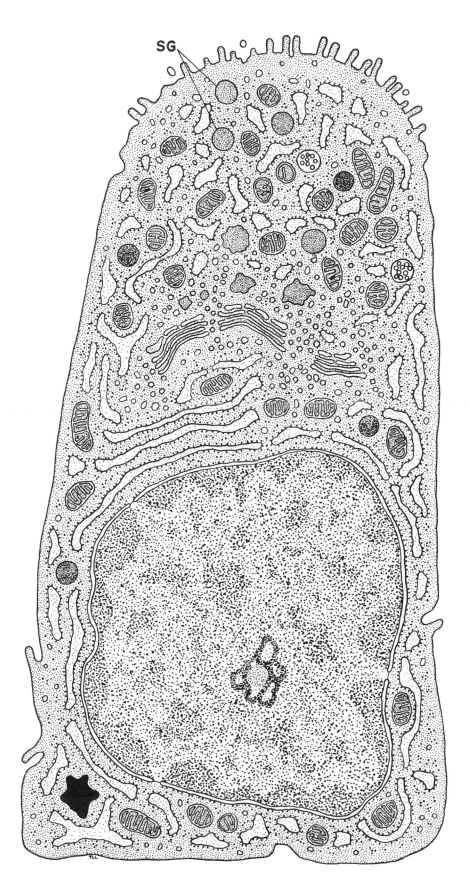

REFERENCES

Brandes, D., 1966. The fine structure and histochemistry of prostatic glands in relation to sex hormones. Int. Rev. Cytol., 20:207–276.

Christensen, A. K., 1965. The fine structure of testicular interstitial cells in guinea pigs. J. Cell Biol., 26:911–935.

Deane, H. W., 1963. Electron microscopic observations on the mouse seminal vesicle. Nat. Cancer Inst. Monogr., 12:63–84.

Fawcett, D. W., 1965. The anatomy of the mammalian spermatozoon with particular reference to the guinea pig. Z. Zellforsch., 67:279–296.

Fawcett, D. W. and R. D. Hollenberg, 1963. Changes in the acrosome of guinea pig spermatozoa during passage through the epididymis. Z. Zellforsch., 60:276–292.

Fawcett, D. W. and S. Ito, 1965. The fine structure of bat spermatozoa. Amer. J. Anat., 116:567–610.

Fawcett, D. W. and D. M. Phillips, 1969. The fine structure and development of the neck region of the mammalian spermatozoon. Anat. Rec., 165:153–184.

Flickinger, C. and D. W. Fawcett, 1967. The junctional specializations of Sertoli cells in the seminiferous epithelium. Anat. Rec., 158:207–222.

Horstmann, E., 1962. Elektronenmikroskopie des menschlichen Nebenhodenepithels. Z. Zellforsch., 57:692–718.

Ladman, A. J., 1967. The fine structure of the ductuli efferentes of the opossum. Anat. Rec., 157:559–576.

Nagano, T., 1966. Some observations on the fine structure of the Sertoli cell in the human testis. Z. Zellforsch., 73:89–106.

Solari, A. J. and L. L. Tres, 1970. The three-dimensional reconstruction of the XY chromosomal pair in human spermatocytes. J. Cell Biol., 45:43–53.

FEMALE REPRODUCTIVE
SYSTEM

FEMALE REPRODUCTIVE SYSTEM

116—OOCYTE

The ovarian follicle consists of a large round oocyte enveloped by follicular cells. A primary oocyte of the human is illustrated here. The nucleus is large and eccentrically located and has a large nucleolus (Nl). Most of the chromatin is dispersed, but there are a few clumps of heterochromatin at the periphery of the nucleus and near the nucleolus. Pores are abundant on the nuclear envelope. A cluster of closely packed, fine, wavy filaments (Fl) is found near or attached to the nuclear envelope. Next to the nucleus is a complicated structure consisting of a cytocentrum, endoplasmic reticulum, Golgi complexes, compound aggregates, annulate lamellae, and mitochondria. This complex corresponds to Balbiani's vitelline body. The cytocentrum is situated in the center of Balbiani's body. The center of the cytocentrum is composed of vesicles, fine filaments, a few mitochondria, and small masses of dense amorphous material. Vesicles (V) are either dispersed or concentrated into aggregates. Some of the dense deposits are arranged in linear arrays that appear to represent a transition between the deposits and coarse, long dense fibers at the periphery of the cytocentrum. In the same region are short, irregular tubules and cisternae of smooth-surfaced endoplasmic reticulum.

Multiple Golgi complexes (G), the compound aggregates (CA), mitochondria, and annulate lamellae (AL) surround the cytocentrum. The Golgi consists of a stack of short lamellae and small vesicles. The compound aggregates are large membrane-limited structures composed of variable amounts of vesicles, particulate material, and vacuoles or droplets resembling lipid; some vesicles envelop a core of dense material. Smaller multivesicular bodies are also present. Mitochondria are densely packed and usually spherical in shape. Sometimes they form a rosette around a mass of dense material. Tubules and cisternae of endoplasmic reticulum occur throughout this area. A few ribosomes are attached to the membranes, and some are free in the cytoplasm.

A single stack of annulate lamellae (AL) is found in the vitelline body, either free or in continuity with the nuclear membrane. The lamellae may be stacked in parallel as shown here, or concentrically arranged. The lamellae are formed by pairs of membranes and are interrupted by large numbers of pores identical in structure to the nuclear pores. Within the pores are cores or plugs of moderately dense material. Some lamellae are continuous with cisternae of endoplasmic reticulum. It has been suggested that the annulate lamellae form by replication of the nuclear envelope and could be involved in nucleocytoplasmic transfer.

Most of the cell organelles are clustered in the paranuclear complex of Balbiani's vitelline body, but smaller numbers of organelles are located in the rest of the cell. The vesicular endoplasmic reticulum is dispersed throughout the cytoplasm. A few of the vesicles contain dense material. Small numbers of mitochondria, Golgi complexes, multivesicular bodies, and dense bodies are found in the cell periphery. Some short microvilli and pinocytotic vesicles are found at the oocyte surface.

Although the oocyte is the only truly totipotent cell in the body, the mature ovum is structurally differentiated and specialized. Many of its specializations are required for subsequent development after fertilization. The immature oocyte, which is less specialized, is engaged in the synthesis of RNA, especially ribosomal RNA. Protein synthesis also occurs and is probably related to the appearance of organelles and yolk protein. The latter is also synthesized by the liver and incorporated into the oocyte from the circulation via the surface pinocytotic vesicles. The mature oocyte is metabolically less active; synthesis of RNA and protein stops but is initiated again by fertilization.

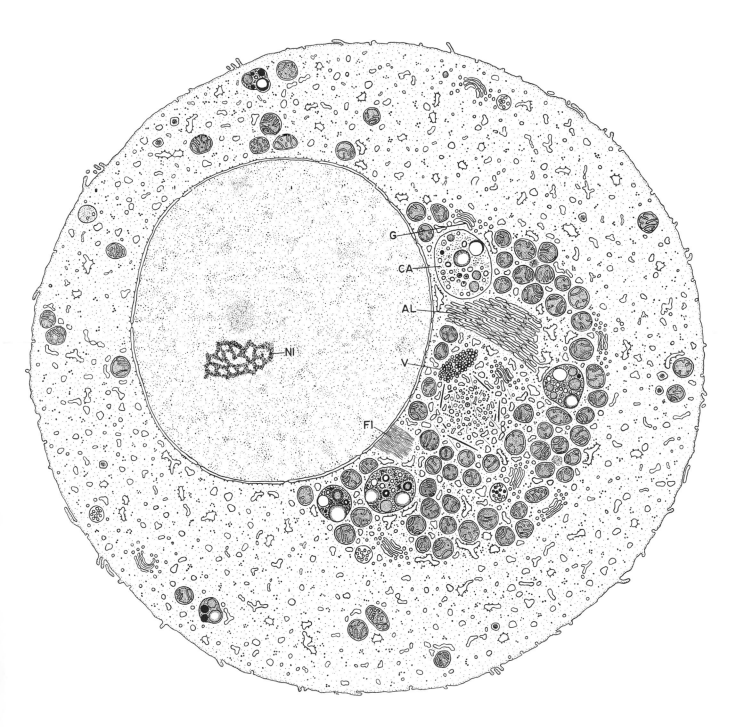

117—FOLLICULAR CELL

The follicular or granulosa cells envelop the primordial oocyte in a single layer but proliferate to form a stratified epithelium in maturing follicles. The follicular cells are closely applied to the oocyte in the early stages but later become separated from it by a mass of homogeneous material comprising the zona pellucida. The zona pellucida is penetrated by cytoplasmic processes extending from the surface of the follicular cells. The cells are cuboidal and have a relatively large nucleus. Cisternae of rough-surfaced endoplasmic reticulum and clusters of ribosomes are relatively abundant. The Golgi apparatus is prominent and usually situated above the nucleus and facing the ovum or antrum. Mitochondria, which are common, are round to oval in shape with lamellar cristae. Finally, a number of lipid droplets (LD), a few dense bodies resembling lysosomes, and larger dense bodies similar to the dense aggregates found in the oocyte or to the lipofuscin pigment bodies found in other cells, occur in the cytoplasm. The granulosa cells secrete mucopolysaccharides into the liquor of the follicular antrum. During the preovulatory phase, they may release an enzyme that depolymerizes the acid mucopolysaccharide to produce a more watery liquor, leading to swelling and rupture of the follicle. Following ovulation, the granulosa cells differentiate into steroid-producing lutein cells.

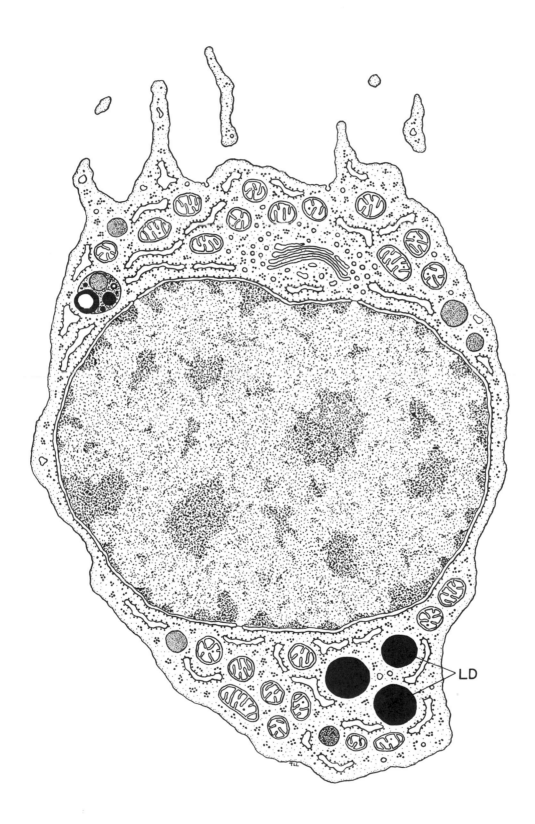

LD

FEMALE REPRODUCTIVE SYSTEM

118—THECA INTERNA CELL

The cells of the theca interna are spindle-shaped or polyhedral cells of connective tissue origin surrounding the ovarian follicle. They are separated from the ovarian follicle by the basement membrane underlying the follicular cells. The cells of the theca interna have a central oval nucleus. The cytoplasm contains a Golgi apparatus, vesicles, short cisternae of rough-surfaced endoplasmic reticulum, free ribosomes, and an occasional lysosome. In these respects, the cells resemble connective tissue cells. They differ, however, by the presence of lipid droplets (LD) and in mitochondrial structure. The mitochondria have some tubular cristae like those of the lutein cells. Because these characteristics are found in steroid-secreting cells, the theca cells are believed to be the source of ovarian estrogen. After ovulation, the theca cells along with the granulosa cells give rise to lutein cells of the corpus luteum, which secrete progesterone.

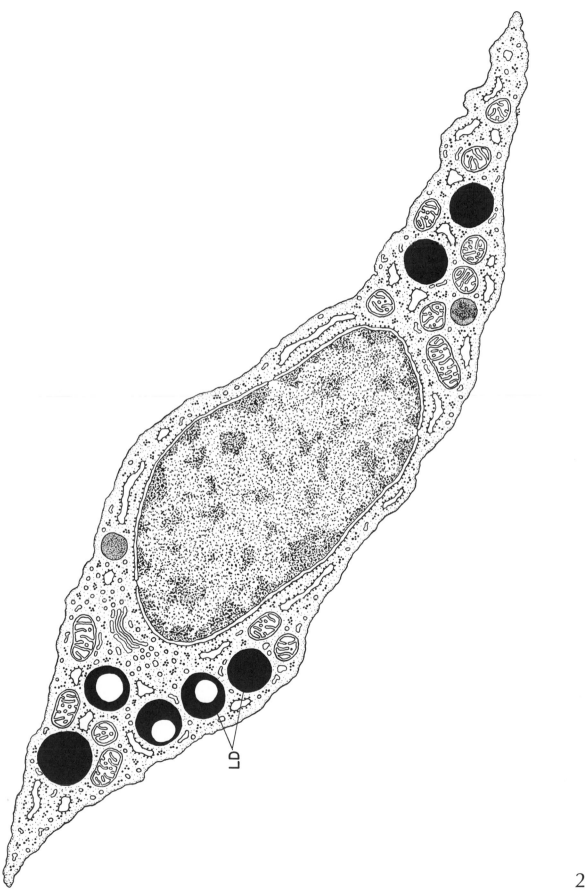

LD

119—LUTEIN CELL

Lutein cells differentiate from the follicular granulosa and theca interna cells after ovulation. Mature active cells are large and polygonal or irregularly shaped with an eccentric nucleus containing a nucleolus. Multiple Golgi complexes are found near the nucleus. These consist of a straight or curved stack of membranous lamellae and many small vesicles. Multivesicular bodies and membrane-bounded dense bodies in a variety of shapes are abundant near the Golgi complexes. Most of the cytoplasm is filled with branching and anastomosing tubules of agranular endoplasmic reticulum (SER). The smooth-surfaced tubules increase in number after ovulation and during pregnancy when steroid synthesis is most intense. They often become densely packed to form regular parallel arrays or concentric whorls. In the periphery of the cell, some parallel stacks of rough-surfaced endoplasmic reticulum (ER) are found and some of these cisternae are continuous with the smooth tubules. Clusters of ribosomes free in the cytoplasm are relatively common. Mitochondria (M) vary from small and spherical to elongated and large and they have vesicular to tubular cristae. Some of the large mitochondria contain masses of dense material in the matrix. A few lipid droplets (LD) are seen in these cells. At the cell surface, clusters of microvilli are situated at irregular intervals. Some blunt projections also occur on the surface and these may indent into and form tight junctions with adjacent cells.

The hormone progesterone is produced by the corpus luteum. Sterol precursors may be stored in the lipid droplets. Mitochondria are involved in pregnenolone formation from cholesterol. The smooth endoplasmic reticulum contains the enzymes responsible for the synthesis of progesterone from pregnenolone. The structure of the luteal cell should be compared to that of other steroid-secreting cells (interstitial cell of the testis, Plate 109; adrenal cortex, Plates 146 to 148).

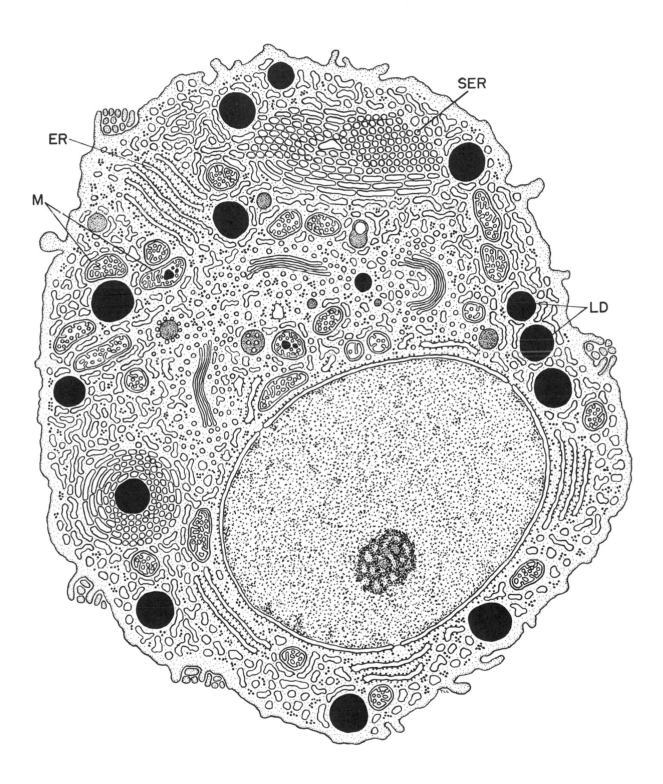

120—FALLOPIAN TUBE: SECRETORY CELL

Three cell types have been described in the epithelium of the oviduct: the secretory cell, the ciliated cell, and an intercalary or "peg" cell. The latter may represent a secretory cell that has expended its granules.

The secretory cells of the oviduct show the structural pattern of cells synthesizing a product for export. The apical region of the cell is filled with secretory granules that have a flocculent content. One component of these granules is mucopolysaccharide. It has been suggested that the secretory product of these cells provides a nutrient material for the ovum, but it may also be involved in maintenance and capacitation (physiologic alterations prerequisite for penetration of the egg) of spermatozoa.

There is a prominent Golgi apparatus above the nucleus and elongated cisternae of rough-surfaced endoplasmic reticulum crowd the basal regions. Nucleoli occur in the nucleus and free ribosomes are common. Mitochondria and a few lysosomes are distributed in the cytoplasm. A number of elongated microvilli project from the apical surface. Like the endometrium, the epithelium of the oviduct undergoes cyclic changes during the menstrual cycle, secretory activity reaching a peak at mid-estrus under the influence of estrogen.

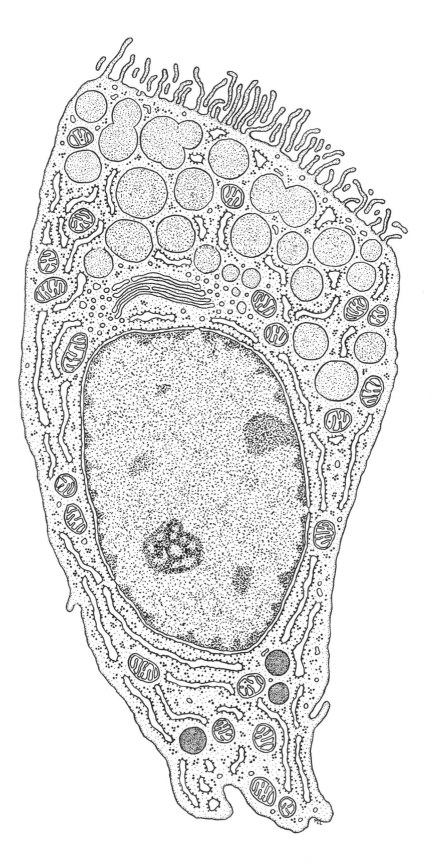

121—FALLOPIAN TUBE: CILIATED CELL

The second major cell type of the oviduct is the ciliated cell. These cells are most abundant in the fimbria and infundibulum and diminish toward the uterus. The cilia in the oviduct beat toward the uterus and might, along with muscular activity of the oviduct, assist movement of the ovum to the uterus. The ciliated cells occur in small groups alternating with clusters of secretory cells. They are columnar with a basal nucleus and, like the secretory cells, are said to vary with the menstrual cycle. The hyaloplasm contains a relatively sparse complement of organelles. Mitochondria are oval to elongated and distributed throughout the cell. A small Golgi apparatus with many small vesicles is located above the nucleus. Cisternae of rough-surfaced endoplasmic reticulum are very rare, but a number of smooth channels and vesicles are seen, especially in the apex of the cell. There is a sparse distribution of ribosomal clusters. A few filaments are present and apically form a terminal web. Some microvilli occur between the cilia on the luminal surface. The cilia terminate in basal bodies from which prominent striated rootlets project deeper into the cytoplasm.

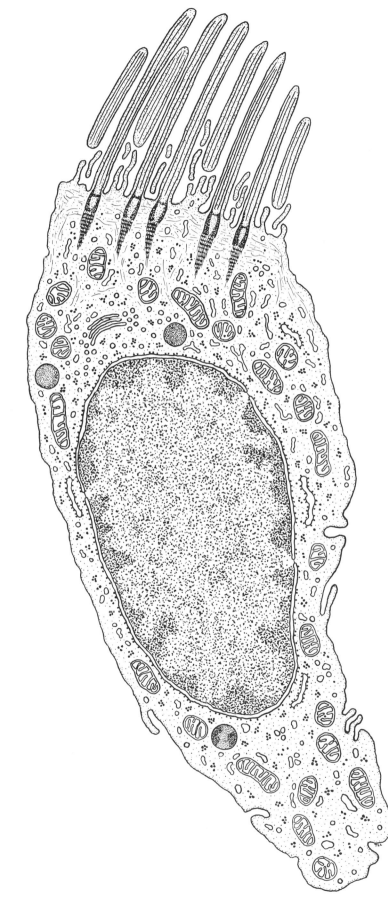

FEMALE REPRODUCTIVE SYSTEM

122—ENDOMETRIAL EPITHELIAL CELL

The glandular endometrial cells in the secretory phase of the menstrual cycle are tall and columnar with an oval nucleus located in the basal region of the cell. The nuclear chromatin is condensed in small peripheral masses. Numerous microvilli are found on the apical cell surface. Larger, blunt cytoplasmic projections containing relatively few organelles also are present. These projections may pinch off into the lumen in a merocrine type of secretion. The lateral plasma membranes interdigitate, especially toward the base of the cell. The Golgi apparatus is situated in a supranuclear position and occupies an extensive area of cytoplasm. Dense secretory granules (SG) occur in variable numbers but become increasingly numerous from the Golgi region to the cell apex. In addition to these, lysosomes, multivesicular bodies, and lipofuscin pigment granules occur. Short to elongated cisternae of rough-surfaced endoplasmic reticulum generally run in the longitudinal axis of the cell. The cisternae are somewhat dilated and tend to occur in greater numbers in the basal regions of the cell. Free ribosomes, many in clusters, are also common. Mitochondria are numerous and vary in shape from spherical to elongated. The cristae extend transversely across the width of the organelle. Glycogen granules are scattered diffusely in the cytoplasm. In some places, especially the base of the cell, glycogen granules (Gly) accumulate in large deposits. Lipid droplets are also present in the base of the cell. Bundles of filaments (Fl) run longitudinally in the cell and form a terminal web area beneath the luminal surface. The glandular secretion is a mucoid fluid rich in nutrients, especially glycogen. Some of the cells in the mucosa are ciliated.

Nucleoli (Nl) are prominent in the early part of the secretory phase; they may be multiple. In addition, the nucleolus forms an unusual structure known as the nucleolar channel system (NCS). This system is located at the periphery of the nucleus adjacent to an invagination of cytoplasm into the nuclear contour. It consists of tubular channels, matrix, and small dense granules. The channels branch and anastomose and appear to be bounded by a membrane. Their content is of low density and has been observed in continuity with the perinuclear space, so that the membrane of the tubular channels is continuous with the inner membrane of the nuclear envelope (NE). The tubules form a basket-like cluster roughly spherical in shape. The interstices between channels are filled with the dense, amorphous matrix material. The channels are surrounded by aggregations of small dense granules. The peripheral region of the system conforms to the usual nucleolar structure. The nuclear envelope separates the nucleolar system from the cytoplasm. However, a nuclear pore (NP) may occur in the nuclear envelope in this region. It has been suggested that the nucleolar channel system represents a morphological basis for a nucleolar-cytoplasmic relationship (Terzakis).

280

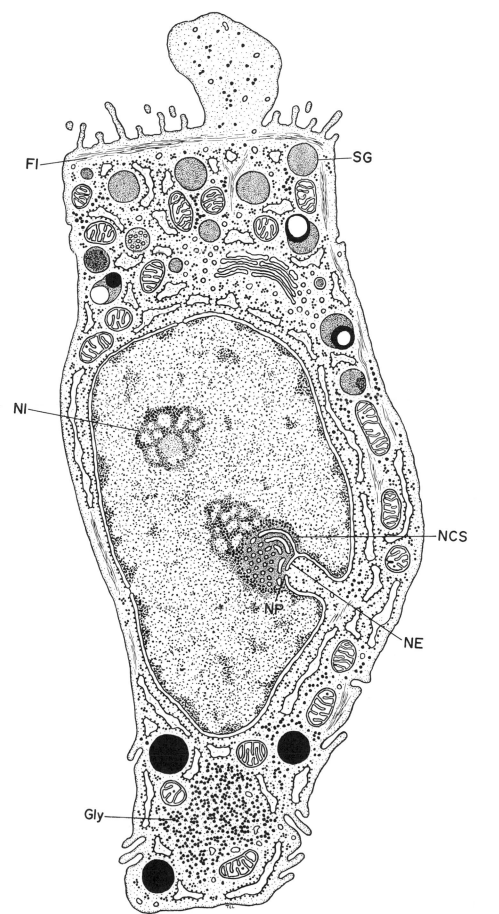

281

123—ENDOMETRIAL GRANULOCYTE

Granular cells populate the endometrium in a wide variety of vertebrate species. This cell can be distinguished from other cells found in the uterus, and it appears to represent a distinct uterine cell type. The granulocyte has been implicated in the secretion of the peptide hormone relaxin.

The cells are spherical with an eccentrically situated, kidney-shaped nucleus. The Golgi apparatus is usually found opposite to the identation in the nucleus. A pair of centrioles occurs near the Golgi complex, and from this region microtubules radiate out to the cell periphery. Most of the granules in the cell are clustered outside the centrosphere region. The granules (Gr) are bounded by a membrane that is either close fitting or separated from the dense material by a clear halo. The clear space underneath the limiting membrane sometimes contains small dense granules or vesicles.

Rough-surfaced endoplasmic reticulum is not prominent; it consists of some elongated cisternae near the nucleus. Smooth-surfaced channels and tubules, on the other hand, are relatively abundant. Free ribosomes are rare and spherical mitochondria are distributed through the cytoplasm. Glycogen granules (Gly) occur often, usually in large masses near the cell surface. The cortex of the cell consists of a zone of filaments devoid of other organelles. The surface itself is irregular and serrated.

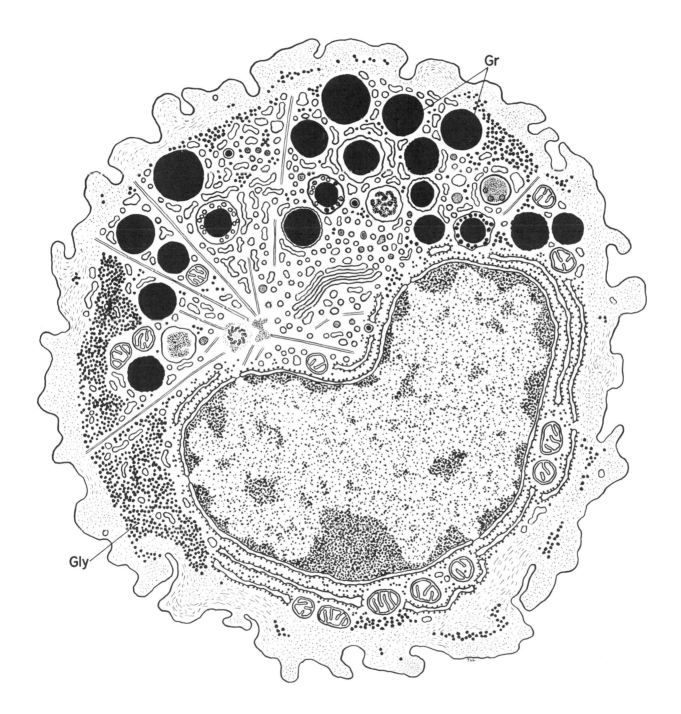

FEMALE REPRODUCTIVE SYSTEM

124 — DECIDUAL CELL

Decidual cells arise from fibroblasts in the uterine stroma adjacent to the site of implantation of the blastocyst. The decidual cells are oval in shape with a central nucleus. The cytoplasm contains clusters of large lipid droplets (LD). Large areas are occupied by glycogen granules (Gly). A number of elongated cisternae of rough-surfaced endoplasmic reticulum containing moderately dense material are seen, and in places these are dilated (ER). Free ribosomes are common also. Bundles of filaments (Fl) run in the cytoplasm among organelles. A small Golgi apparatus occurs near the nucleus, and mitochondria and lysosomes are scattered throughout the cell. A few small vesicles are associated with the cell surface. A thin ectoplasmic zone contains filaments but few organelles. The surface is somewhat irregular with a few villi or blunt projections. Junctional complexes are formed between adjacent cells. During decidual transformation the fibroblasts assume some epithelioid characteristics. In addition, they appear to be capable of protein synthesis and to be engaged in the storage of lipid and glycogen.

DECIDUAL CELL

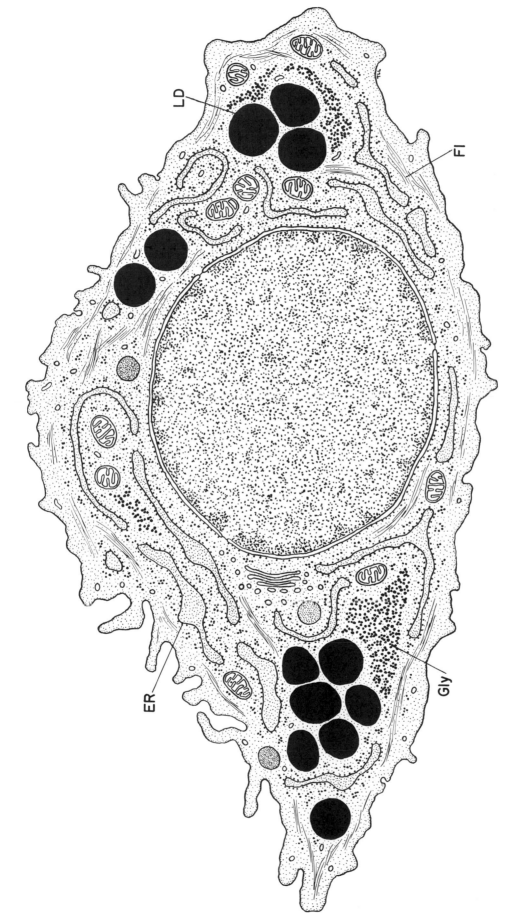

285

125 — MAMMARY GLAND CELL

The glandular epithelial cells of the mammary gland are cuboidal. Scattered microvilli occur on the free surface. The lateral borders are bound together by junctional complexes and interdigitate, while a few villous processes occur basally. The nucleus is situated toward the base of the cell and contains a nucleolus. Two particulate components are found in the milk of the mammary gland: small dense protein granules (PG) and large lipid droplets (LD).

The protein granules are elaborated and discharged in a manner similar to that in other cells secreting protein. The synthetic apparatus consists of many free ribosomes and an extensive system of rough-surfaced endoplasmic reticulum that is most prominent in the base of the cell where the lamellae are stacked in parallel. A large Golgi apparatus is located above the nucleus. Associated with the Golgi are some vacuoles containing fibrillar or particulate material that condenses into a central core or granule. Some of the vacuoles are small and contain one granule, but toward the apex the vacuoles become progressively larger and contain more dense protein granules. The vacuoles fuse with the surface membrane and liberate their contents intact into the lumen.

Fat droplets are released into the milk in a different manner. They are found throughout the cell but are largest near the apex. Here, they protrude into the lumen and appear to pinch off from the cell proper, along with a small rim of cytoplasm that contains the organelles. Other cytoplasmic structures are large mitochondria with closely packed lamellar cristae, lysosomes, and a small number of smooth membranous tubules and vesicles.

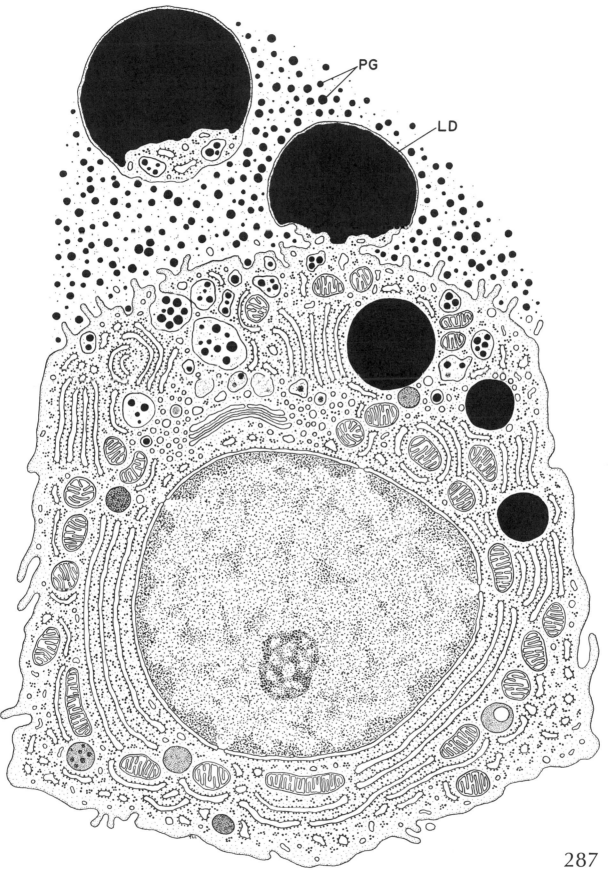

REFERENCES

Adams, E. C. and A. T. Hertig, 1969. Studies on the human corpus luteum. I. Observations on the ultrastructure of development and regression of the luteal cells during the menstrual cycle. J. Cell. Biol., 41:696–715.

Bjersing, L., 1967. On the ultrastructure of granulosa lutein cells in porcine corpus luteum. With special reference to endoplasmic reticulum and steroid hormone synthesis. Z. Zellforsch., 82:187–211.

Cardell, R. R., F. L. Hisaw, and A. B. Dawson, 1969. The fine structure of granular cells in the uterine endometrium of the Rhesus monkey (*Macaca mulatta*) with a discussion of the possible function of these cells in relaxin secretion. Amer. J. Anat., 124:307–340.

Cavazos, F., J. A. Green, D. G. Hall, and F. V. Lucas, 1967. Ultrastructure of the human endometrial glandular cell during the menstrual cycle. Amer. J. Obstet. Gynec., 99:833–854.

Fredricsson, B. and N. Björkman, 1962. Studies on the ultrastructure of the human oviduct epithelium in different functional states. Z. Zellforsch., 58:387–402.

Helminen, H. J. and J. L. E. Ericsson, 1968. Studies on mammary gland involution. I. On the ultrastructure of the lactating mammary gland. J. Ultrastruct. Res., 25:193–213.

Hertig, A. T., 1968. The primary human oocyte: Some observations on the fine structure of Balbiani s vitelline body and the origin of the annulate lamellae. Amer. J. Anat., 122:107–138.

Hertig, A. T. and E. C. Adams, 1967. Studies on the human oocyte and its follicle. I. Ultrastructural and histochemical observations on the primordial follicle stage. J. Cell. Biol., 34:647–675.

Jollie, W. P. and S. A. Bencosme, 1965. Electron microscopic observations on primary decidua formation in the rat. Amer. J. Anat., 116:217–236.

Kurosumi, K., Y. Kobayashi, and N. Baba, 1968. The fine structure of mammary glands of lactating rats, with special reference to the apocrine secretion. Exp. Cell Res., 50:177–192.

Terzakis, J. A., 1965. The nucleolar channel system of human endometrium. J. Cell Biol., 27:293–304.

Wynn, R. M. and J. A. Harris, 1967. Ultrastructural cyclic changes in the human endometrium. I. Normal preovulatory phase. Fertil. Steril., 18:632–648.

Wynn, R. M. and R. S. Woolley, 1967. Ultrastructural cyclic changes in the human endometrium. II. Normal postovulatory phase. Fertil. Steril., 18:721–738.

EMBRYONIC TISSUES

EMBRYONIC TISSUES

126—BLASTOCYST: TROPHOBLAST CELL

The blastocyst is a sphere with an outer layer, or trophoblast, enclosing a hollow cavity and an inner cell mass. The inner cell mass is applied to the inner surface of the trophoblast at one pole of the blastocyst. During implantation, the blastocyst becomes apposed to the uterine epithelium, adheres to it, and then penetrates the endometrium. The outer trophoblast layer gives rise to the cytotrophoblast (Plate 128) and the syncytiotrophoblast (Plate 129), which together constitute the fetal portion of the placenta.

Prior to implantation, the trophoblast is a single layer of cuboidal cells joined laterally by junctional complexes. The nucleus is central in position and has a prominent nucleolus. The outer or apical surface has a number of microvilli, while the basal surface is relatively smooth in contour. The cytoplasmic structure is that of an undifferentiated cell. Most conspicuous is the large number of free ribosomes (R) that tend to occur in clusters. Elements of rough-surfaced endoplasmic reticulum, on the other hand, are few in number and consist only of isolated cisternae. A small Golgi apparatus is found, but there is no evidence of secretory activity. Mitochondria are generally spherical and have a dense matrix. Lipid droplets and vacuoles that contain particulate material are seen occasionally.

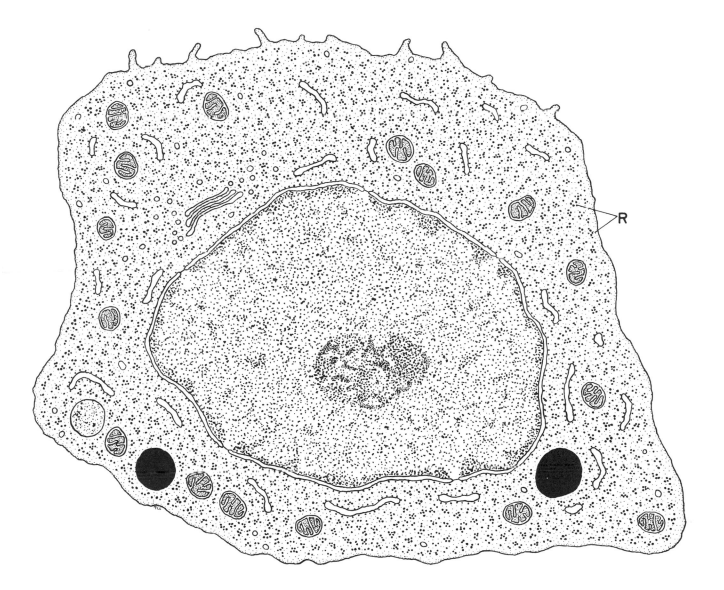

EMBRYONIC TISSUES

127—BLASTOCYST: INNER CELL MASS CELL

The cells of the inner cell mass are destined to form the embryo proper. They are closely packed and have a central nucleus containing a prominent nucleolus. The chromatin is not highly condensed. These cells differ from those of the trophoblast in two respects. First, cytoplasmic ribosomes are not as abundant in these cells, although there are still a large number. Most of these ribosomes occur in clusters. Second, cells of the inner cell mass contain a variety of vacuoles (Vac) with a heterogeneous content. These structures are composed of a membrane that encloses particulate and lamellar components. Multivesicular bodies and lipid droplets (LD) are found as well. Many of these vacuoles may be related to segregation and digestion of materials necessary for nutrition of the cells, and they could be derived from yolk bodies that were present earlier. In other respects, the cells are unspecialized. A Golgi apparatus occurs, but endoplasmic reticulum is not prominent. Mitochondria are like those of the trophoblast.

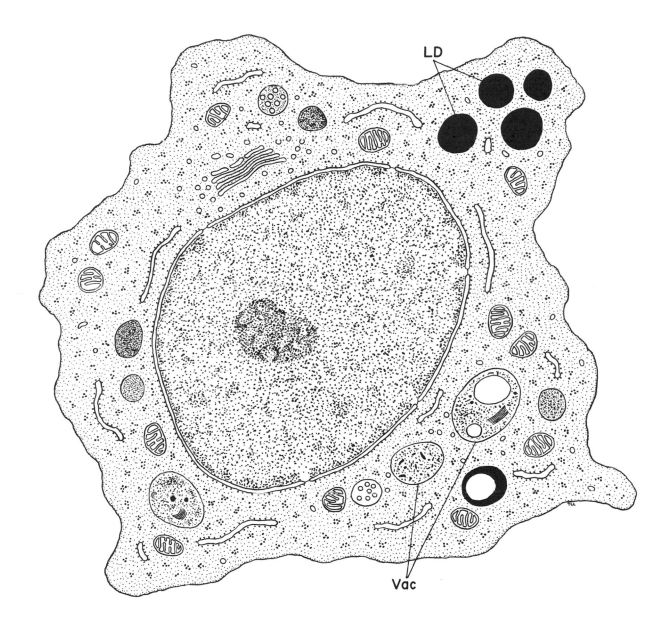

LD

Vac

EMBRYONIC TISSUES

128—CYTOTROPHOBLAST CELL

The villi of the early placenta have a connective tissue core that contains fetal capillaries and a covering of trophoblast that consists of two cell layers. The inner cytotrophoblast, or Langhans layer, rests on a basement lamina and is composed of discrete cells. The outer layer is the syncytial trophoblast, in which lateral cell borders are absent.

The trophoblast layer undergoes changes with increasing placental age. Initially, the cytotrophoblast forms a complete layer beneath the syncytial trophoblast. Later, this layer becomes discontinuous until at term only a few isolated cells persist. At the same time, the cells show cytological changes and progress from relatively undifferentiated cells to specialized cells containing numerous filaments. The differentiated cells are somewhat flattened. The nucleus is central in position and contains a nucleolus. Elongated cisternae of rough-surfaced endoplasmic reticulum (ER) are arranged circumferentially around the nucleus. The cisternae are somewhat dilated and contain moderately dense material. Free ribosomes, as well as glycogen granules, are abundant in the cytoplasm. A Golgi apparatus occurs near the nucleus and has a large number of small vesicles and a few dense granules associated with it. Lysosomes, smooth membranes, and, occasionally, lipid droplets are found in the cytoplasm. Mitochondria are small and usually spherical but occur in relatively large numbers. The cytoplasmic filaments (Fl) run in bundles between the cell organelles. The cytotrophoblast cells are connected to each other and to the overlying cells by desmosomes. Laterally, some irregular microvilli project into the intercellular space.

Several functions have been attributed to the cytotrophoblast cells. They undergo mitosis and contribute to the formation of the syncytial trophoblast. They have similarities with connective tissue cells and may give rise to the fibrin and fibrinoid material that accumulates in the intercellular spaces late in pregnancy. It has also been suggested that they function in erosion of the decidua by lytic activity and that they secrete the glycoprotein hormone chorionic gonadotropin; however, they bear few cytological resemblances to enzyme-secreting or hormone-forming cells.

294

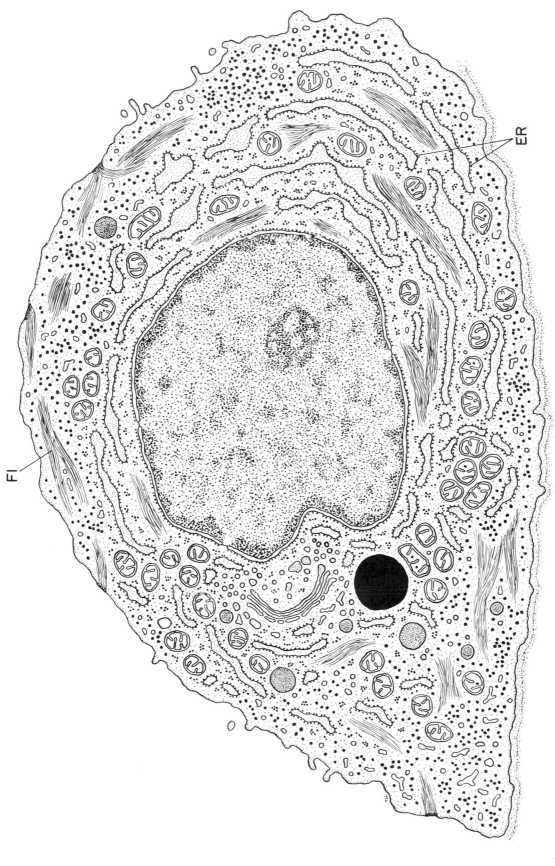

ER

FI

129—SYNCYTIOTROPHOBLAST

The syncytial trophoblast forms the outer covering of the placental villi and appears to represent one of the few examples of a true syncytium. The electron microscope reveals no lateral plasma membranes between nuclei of this layer. The syncytial trophoblast becomes progressively thinned out as the placenta ages, decreasing the distance between fetal and maternal circulation. In most places in the term placenta, the syncytial trophoblast forms the only layer of the trophoblast and rests on the basement membrane, the cytotrophoblast having largely disappeared by this time.

Many microvilli (Mv) project from the cell surface into the maternal blood space. Small, pinocytotic vesicles (PV) are common at the bases of the microvilli. The basal plasma membrane bears a number of short infoldings and processes. The nucleus is roughly equidistant between the basal and apical membranes of the layer. The endoplasmic reticulum consists of many short, ribosome-studded cisternae. Small Golgi complexes are common and are seen on either side of the nucleus. Moderately dense, small, membrane-bounded granules (Gr) appear to originate in the Golgi regions. Mitochondria are small and not complex in internal structure. Bundles of filaments occur in the cytoplasm but are not nearly as extensive as in cytotrophoblast cells.

The syncytial trophoblast serves as the primary organ of interchange of substances between the fetal and maternal blood streams. Oxygen and all other substances necessary for the nutrition of the fetus must cross this layer. The microvilli greatly increase the absorptive surface of the syncytium and probably facilitate uptake. The small invaginations of the surface plasma membrane may be involved in either uptake or expulsion of substances. Waste products from the fetus, such as creatinine, urea, and carbon dioxide, pass in the opposite direction across the placenta. The basal infoldings may play a role in excretion in a manner similar to that occurring in the convoluted tubule cells of the kidney and in other epithelial tissues engaged in water transport. The placenta also serves as a barrier preventing many substances, especially particulate ones like maternal erythrocytes and bacteria, from entering the fetal circulation. Finally, the syncytial trophoblast is thought to be the source of the steroid hormones estrogen and progesterone, which are secreted by the placenta.

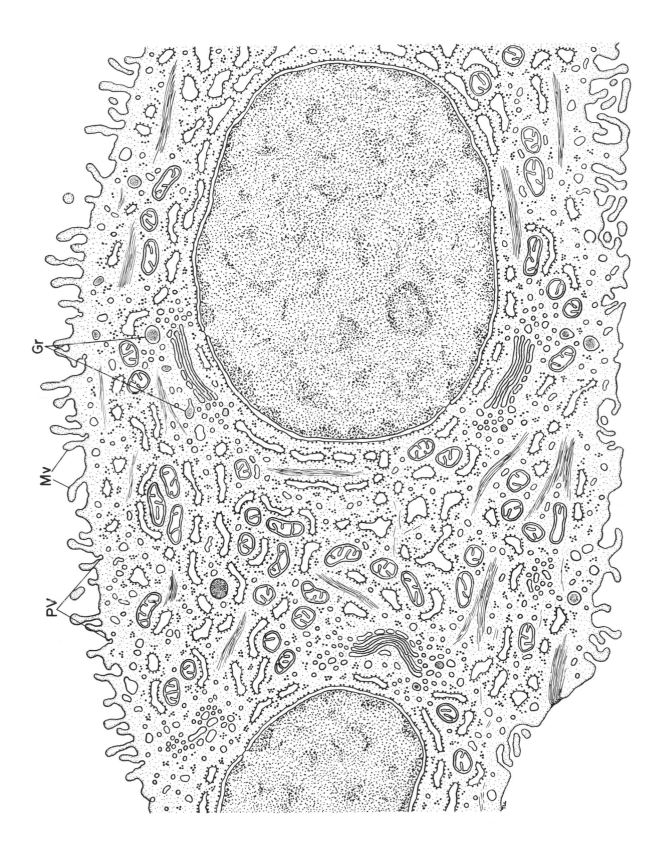

EMBRYONIC TISSUES

130—AMNIOTIC EPITHELIAL CELL

The amnion epithelium is formed by a single layer of cuboidal cells of embryonic origin that face the amniotic cavity. Perhaps the most striking feature of these cells is the large surface area formed by villi and by infoldings of the surface membrane. The rounded surface of the cells that faces the amniotic cavity is covered with pleomorphic microvilli (Mv). Laterally, villi project into intercellular canals and cytoplasmic processes interdigitate with those of adjacent cells. The basal membrane, too, is highly irregular with numerous infoldings and extensions. Desmosomes (*maculae adherentes*, MA) are common along the lateral borders. Vesicles and small vacuoles are continuous with the membranes of all surfaces but seem more abundant along the basal processes and lateral intercellular canals.

The nucleus is indented, located toward the base, and may contain a nucleolus. The Golgi zone tends to be situated above the nucleus and is elaborate in structure with many small vesicles associated with it. The rough-surfaced endoplasmic reticulum (ER) is also extensive and it consists of dilated cisternae containing moderately dense material. Relatively small mitochondria, ribosomes, and a few lipid droplets are distributed in the cytoplasm. Bundles of filaments extend between the organelles and are especially common in the apex and base. In some cells, the Golgi apparatus and endoplasmic reticulum are not as extensive, but cytoplasmic filaments are even more abundant.

The cytological specializations indicate that the amnion cells are engaged in the transport of fluids, that they are synthesizing proteins, and that they form a coherent sheet of tightly bound cells. That is, the amplification of the surface membranes suggests a cell active in fluid transport. The amniotic fluid is probably initially formed by the amnion cells, but there is in addition a constant turnover of the fluid. The amnion becomes fused with the chorion, so that materials in the amniotic fluid could pass across the chorioamnion into the maternal circulation; transport could also occur in the opposite direction. The cells of the amnion epithelium also appear to be synthesizing material and, although the nature of the secretory product is unknown, relatively large amounts of γ-globulin occur in the amniotic fluid. Finally, the amnion serves as a protective cover and is probably subjected to considerable stress and pressure. The numerous desmosomes between adjacent cells, as well as the system of cytoplasmic fibrils, may serve to join these cells into a coherent, tightly bound sheet.

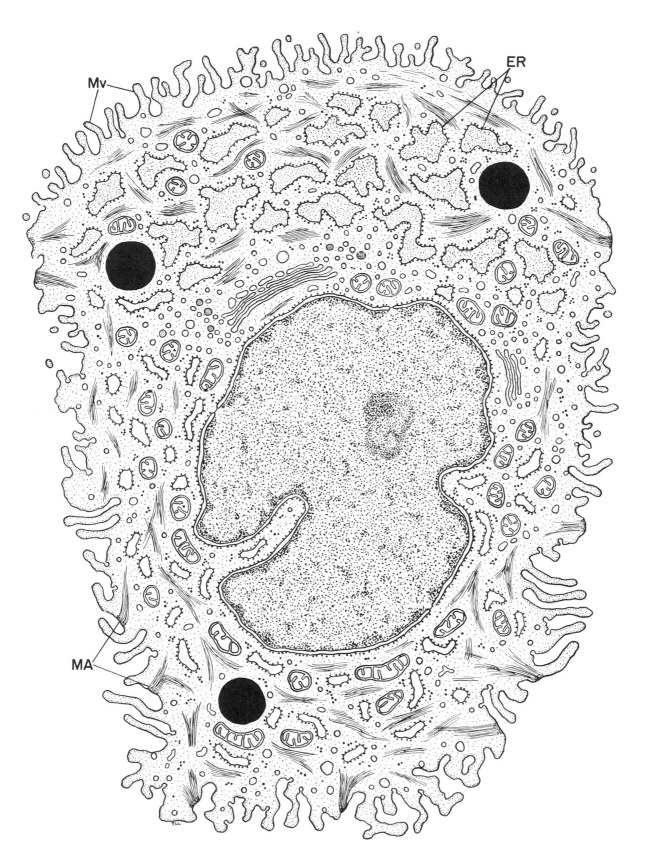

131—YOLK SAC EPITHELIAL CELL

Cells of the visceral yolk sac endoderm are cuboidal to columnar in shape with a dome-like apex. The nucleus, which is indented, is situated at the base of the cell. It contains masses of condensed chromatin and a nucleolus. The free cell surface bears numerous microvilli. Typical junctional complexes (not shown) are formed between adjacent cells. The lateral cell border is irregular with folds and interdigitations. Basally, there are a few shallow folds and undulations.

At the free surface, saccular invaginations of the plasma membrane are seen. These are lined by a layer of filamentous material and have a bristle coat on their cytoplasmic surface (coated vesicles). The apical cytoplasm contains large numbers of small apical pits (AP) and vesicles (AV) with flocculent or dense material. In the same region are short tubules or channels (Ch) containing moderately dense material. Somewhat deeper in the cytoplasm are larger vacuoles (Vac) that enclose flocculent material. These structures may be related to absorptive activity. The apical invaginations and vesicles may be involved in selective adsorption of proteins (see also Plates 12, 19, 99, 111, 112). These pinch off from the surface to lie as vesicles free in the apical cytoplasm. The tubules could represent vesicles or saccules from which the fluid content has been withdrawn. Fusion of vesicles would lead to the formation of larger vacuoles. Some of the proteins taken up by the epithelial cells may be degraded by lysosomes. Other proteins, such as antibodies, might pass through the cells to reach the fetus.

The supranuclear region of the cell contains some large granules with a dense content. It is not certain whether these are derived from the large vacuoles or represent a type of lysosome. Similar granules occur in the base of the cell. A Golgi apparatus occurs above the nucleus. There are a few scattered rough-surfaced cisternae and a number of profiles of smooth endoplasmic reticulum in the cytoplasm. Stacks of rough-surfaced cisternae are encountered in the basal zone. A few lipid droplets occur in these cells and glycogen granules are common. Mitochondria are oval and have a dense matrix containing opaque granules. Some of the mitochondria near the nucleus are relatively large. Filaments occupy the cores of the villi and apical cytoplasm and a few microtubules are seen.

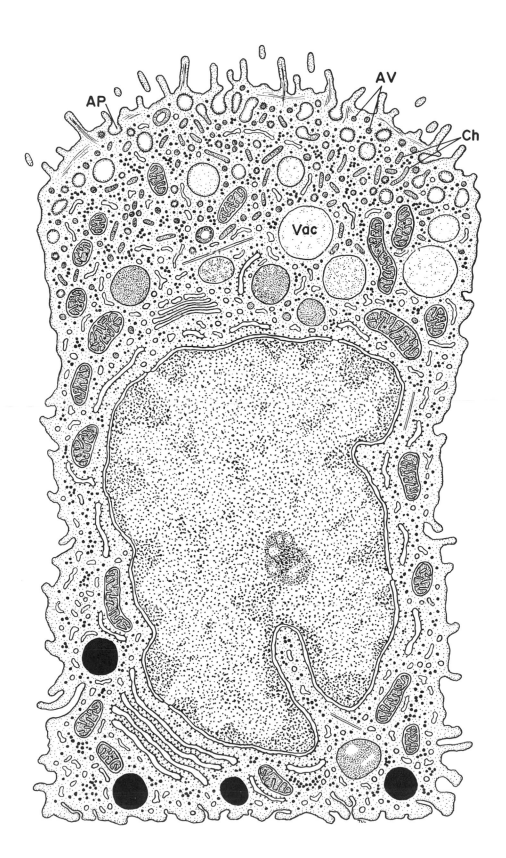

EMBRYONIC TISSUES

132—UMBILICAL CORD: MUCOUS CONNECTIVE TISSUE CELL

Wharton's jelly of the umbilical cord is a loose, mucous connective tissue composed of mesenchymal cells widely separated by intercellular material. The surface of the umbilical cord is lined by amniotic epithelium (Plate 130). The mesenchymal cells are specialized fibroblasts with a stellate or spindle shape. The most conspicuous feature of these cells is their extensive rough-surfaced endoplasmic reticulum. Cisternae occupy much of the cytoplasm; they are elongated and slightly dilated and contain moderately dense material. Free ribosomes occurring in clusters are also extremely common. Other organelles are not as prominent or numerous. A Golgi apparatus is found near the centrally situated nucleus. A few small dense granules and some mitochondria with a dense matrix are distributed throughout the cell. The surface is somewhat irregular and has a small number of vesicles associated with it. These cells appear to be actively synthesizing proteins and are responsible for the production of the intercellular materials, which include collagen and mucopolysaccharide.

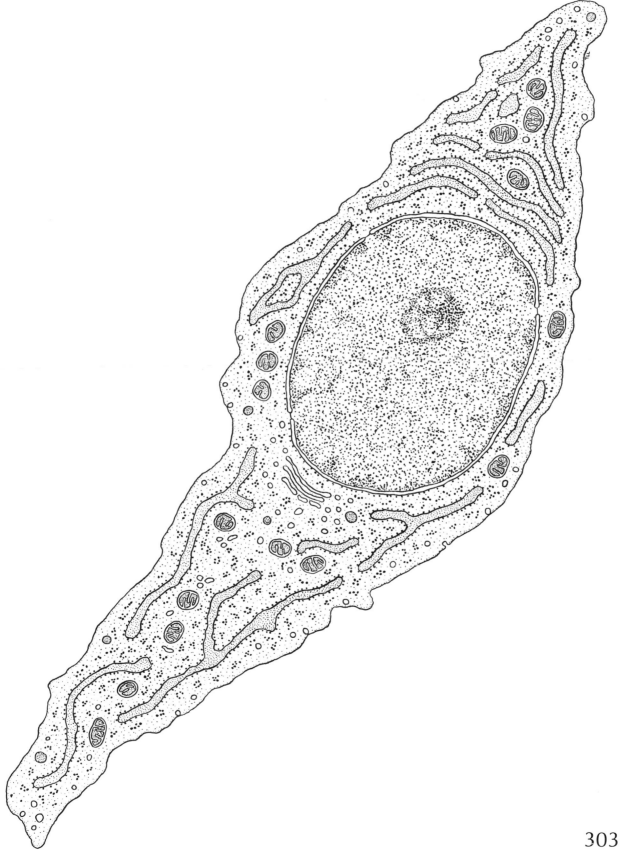

REFERENCES

Enders, A. C., 1965. A comparative study of the fine structure of the trophoblast in several hemochorial placentas. Amer. J. Anat., 116:29–68.

Enders, A. C., 1968. Fine structure of anchoring villi of the human placenta. Amer. J. Anat., 122:419–452.

Enders, A. C. and S. Schlafke, 1967. A morphological analysis of the early implantation stages in the rat. Amer. J. Anat., 120:185–226.

Enders, A. C. and S. Schlafke, 1969. Cytological aspects of trophoblast-uterine interaction in early implantation. Amer. J. Anat., 125:1–30.

King, B. F. and A. C. Enders, 1970. The fine structure of the guinea pig visceral yolk sac placenta. Amer. J. Anat., 127:397–414.

Leeson, C. R. and T. S. Leeson, 1965. The fine structure of the rat umbilical cord at various times of gestation. Anat. Rec., 151:183–198.

Potts, D. M. and I. B. Wilson, 1967. The preimplantation conceptus of the mouse at 90 hours *post coitum.* J. Anat., 102:1–11.

Thomas, C. E., 1965. The ultrastructure of human amnion epithelium. J. Ultrastruct. Res., 13:65–84.

ENDOCRINE SYSTEM

133—PITUITARY: SOMATOTROPH

The pituitary gland or hypophysis consists of the adenohypophysis, which arises as a dorsal outpocketing (Rathke s pouch) of the lining of the oral cavity, and the neurohypophysis, which develops as a downgrowth from the floor of the diencephalon. The adenohypophysis is made up of the pars distalis or anterior lobe, the pars tuberalis or pars infundibularis, and the pars intermedia. The subdivisions of the neurohypophysis are the pars nervosa or infundibular process, and the infundibulum (infundibular stalk and median eminence of the tuber cinereum).

The cells of the pars distalis or anterior lobe of the pituitary have been subdivided into acidophilic, basophilic, and chromophobic cells on the bases of the staining affinities of their specific granules. Acidophils (alpha cells) are believed to secrete growth hormone (somatotropin, STH) and lactogenic hormone (LTH, mammotropin, or prolactin). Basophils (beta cells) are thought to secrete follicle-stimulating hormone (FSH), luteinizing hormone (LH, in the male called interstitial cell-stimulating hormone, ICSH), and thyroid-stimulating hormone (TSH, thyrotropin). The cell from which adrenocorticotropic hormone (ACTH, corticotropin) originates is uncertain, although some evidence points to the basophil. The chromophobes lack secretory granules and thus do not stain in the manner of the chromophilic cells. They have been regarded as chromophilic cells that have expended their granules, or as a reserve population capable of giving rise to other cell types. With the electron microscope, a separate cell type has been attributed to each of the six hormones.

Somatotrophs, producers of somatotropic hormone (STH), contain dense secretory granules (SG) with a diameter of 300 to 350 mμ. The granules are abundant and fill much of the cell. The cell has a central nucleus near which is a large Golgi apparatus (G). Small granules in the Golgi zone may represent immature secretory granules. Mitochondria and flattened cisternae of rough-surfaced endoplasmic reticulum are found in the cytoplasm between granules. Some ribosomes are free in the cytoplasm. Secretory product has been observed lying within small concavities of the plasma membrane outside the cell. This dense material is not bounded by a membrane and lies between the plasma membrane of the cell and the basal lamina. These images suggest release of secretory product from the cell by a process of reverse pinocytosis. STH is a protein with a molecular weight of 21,000. It stimulates body growth after birth, especially growth of the epiphyseal cartilages.

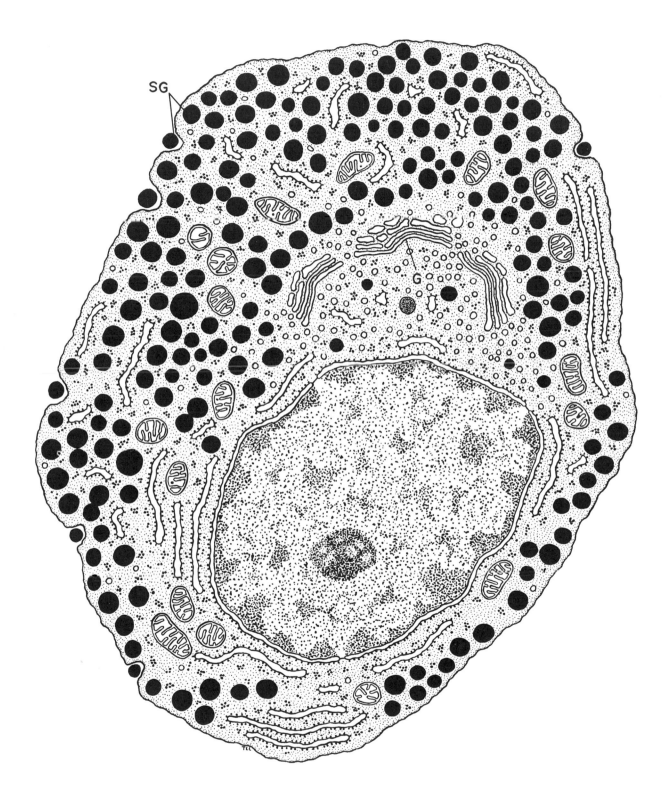

134—PITUITARY: MAMMOTROPH

Mammotrophs are identified on the basis of their content of large secretory granules (SG). The dense granules are 600 to 900 mμ in diameter and vary in shape from spherical to polymorphous. The number and size of the mammotrophs increases during pregnancy and during lactation, supporting the belief that these cells secrete lactogenic hormone (LTH, mammotropin, also called prolactin). This hormone is a protein with a molecular weight of about 25,000. It stimulates lactation by the mammary gland and, at least in some animals, it has a luteotropic effect in promoting secretion of progesterone by the corpus luteum.

Actively secreting cells have a large juxtanuclear Golgi complex. The Golgi complex consists of curved cisternae and many smooth vesicles. The inner cisternae, more flattened than the outer, frequently contain small masses of dense secretory material. A number of immature secretory granules are found in the core of cytoplasm circumscribed by the Golgi cisternae. The immature granules are polymorphous and consist of aggregates of smaller granules. Mature granules are rounded or ovoid and are most common between the Golgi complex and the endoplasmic reticulum and along the cell membrane.

The rough-surfaced endoplasmic reticulum consists of long, flattened cisternae running parallel to the cell membrane. A number of irregular profiles occur near the Golgi apparatus. Some of these correspond to junction elements between the endoplasmic reticulum and Golgi complex. Ribosomes, mitochondria, and lysosomes (Ly) are also found in the cytoplasm. Lysosomes and their derivatives are much more prominent when secretion is suppressed, and they are believed to function in the regulation of the secretory process by degrading secretory granules. When secretory activity is high, the cells have abundant rough-surfaced endoplasmic reticulum, a large Golgi apparatus, and many forming secretory granules. However, when secretion is suppressed, secretory granules and cytoplasmic constituents are sequestered into lytic bodies and progressively degraded.

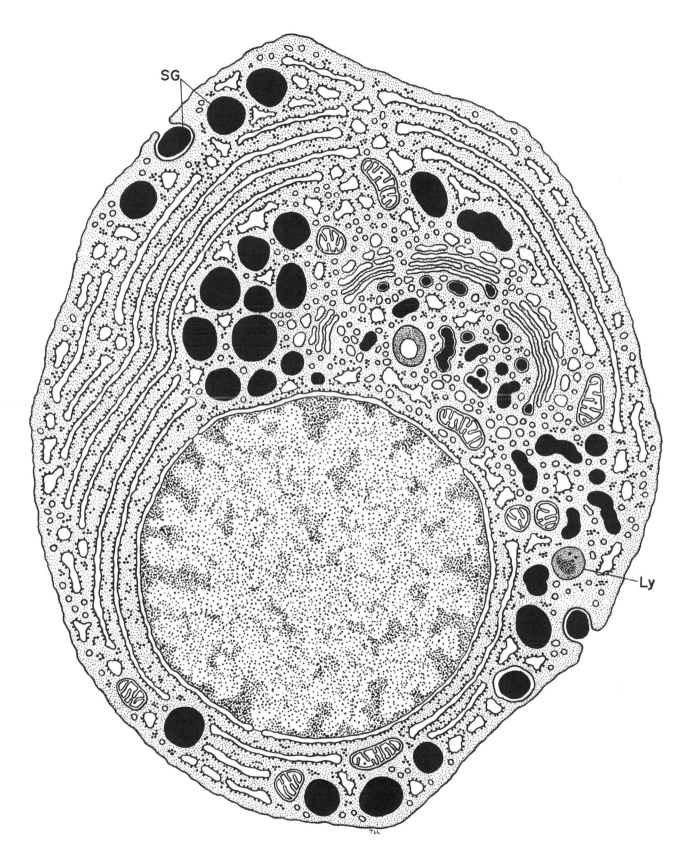

ENDOCRINE SYSTEM

135—PITUITARY: FOLLICLE-STIMULATING HORMONE (FSH) GONADOTROPH

The gonadotroph that produces follicle-stimulating hormone (FSH) has a large round cell body. A well-developed Golgi apparatus (G) occurs beside the nucleus. The Golgi complex consists of membranous lamellae and many small vesicles and it has immature secretory granules associated with it. Dense secretory granules (SG) are abundant in the cytoplasm and are usually about 200 mμ in diameter. The rough-surfaced endoplasmic reticulum (ER) is composed of distended vesicular elements. These dilated cisternae contain flocculent material of medium density. Some ribosomes are free in the cytoplasm. Mitochondria and a few large dense granules resembling lysosomes are scattered through the cytoplasm. FSH is a glycoprotein hormone with a molecular weight between 30,000 and 67,000. In the female, it stimulates growth of ovarian follicles and, in the male, it stimulates the seminiferous epithelium of the testis. Synthesis of the protein hormones presumably takes place on ribosomes of the endoplasmic reticulum via the same mechanism that occurs in the pancreatic acinar cell. However, in the case of the cells producing glycoproteins, the Golgi complex may play a role in the synthesis of the carbohydrate moiety of the carbohydrate-protein complexes, as well as concentrating the final secretory product.

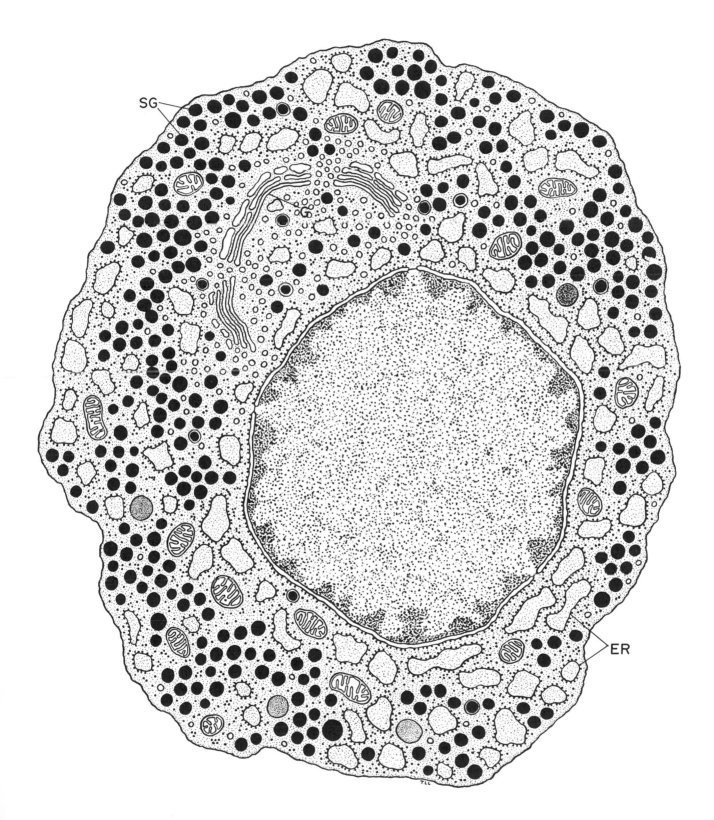

136—PITUITARY: LUTEINIZING HORMONE (LH) GONADOTROPH

The second type of gonadotroph secretes luteinizing hormone (LH), also called interstitial cell-stimulating hormone (ICSH) in the male. The secretory product is contained in dense granules that are more uniform in size and slightly larger, about 250 mμ in diameter, than those of the FSH gonadotroph. The granules tend to accumulate in one pole of the cell and near the cell periphery. The Golgi apparatus is not as extensive and the endoplasmic reticulum is composed of some flattened rough-surfaced cisternae in contrast to the dilated sacs of the FSH gonadotroph. The other cytoplasmic structures are mitochondria and free ribosomes. The round nucleus is central or slightly eccentric in position.

LH is a glycoprotein like FSH and has a molecular weight of 26,000. LH is necessary for the conversion of a ruptured follicle into a corpus luteum after the follicles have been stimulated by FSH. In the male, the hormone (ICSH) stimulates the interstitial cells of the testis to secrete androgen.

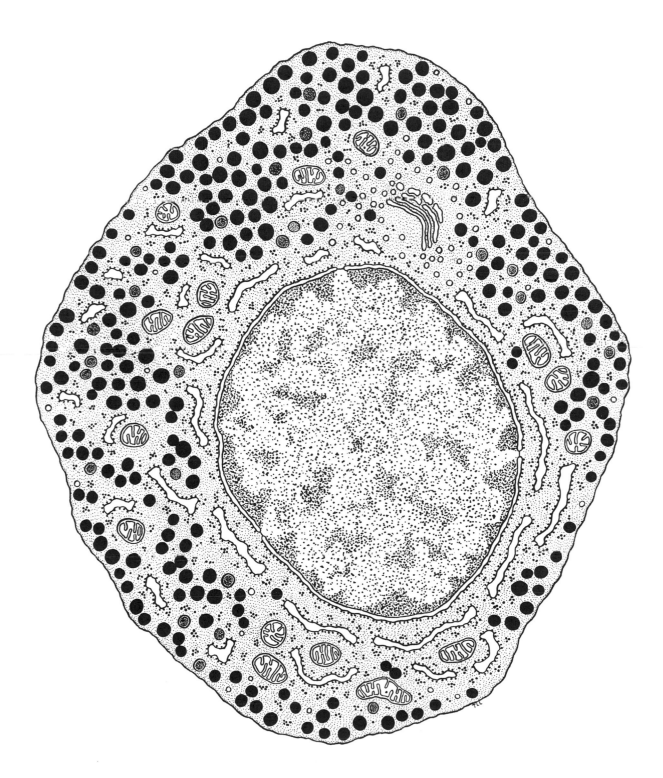

137—PITUITARY: THYROTROPH

Thyrotrophs, the producers of thyroid-stimulating hormone (TSH), are smaller than the other cell types in the anterior lobe and are often angular or irregular in shape. The nucleus is similarly flattened. The cells are easily identified by their granules, which are also smaller than the other types, being only 120 to 150 mμ in diameter. The granules are characteristically aligned one row deep adjacent to the plasma membrane. A smaller number of granules occurs deeper in the cytoplasm and in association with the Golgi apparatus. Other cytoplasmic structures are mitochondria, flattened cisternae of rough-surfaced endoplasmic reticulum, and ribosomes. TSH is another glycoprotein hormone and has a molecular weight between 10,000 and 30,000. It stimulates the secretory activity of the follicular cells of the thyroid.

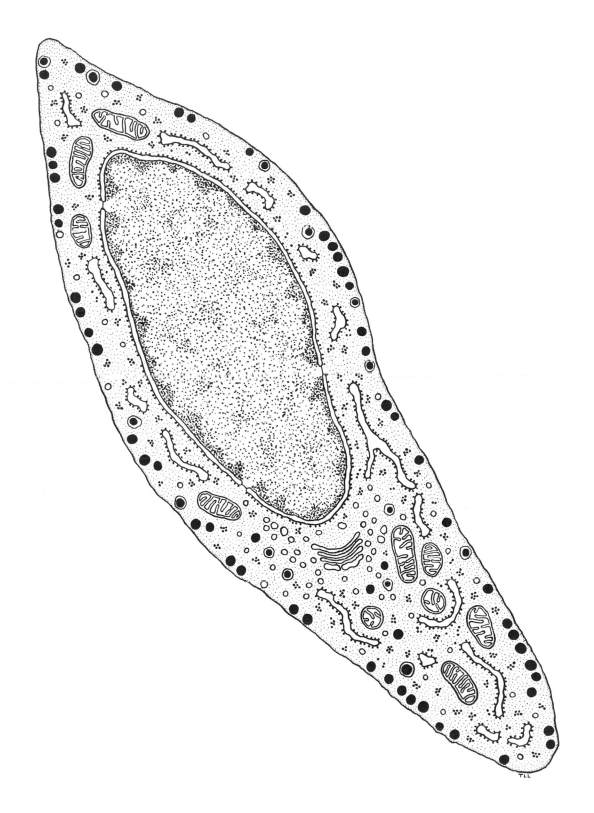

138—PITUITARY: CORTICOTROPH

The corticotroph produces adrenocorticotrophic hormone (ACTH). It is large and irregular in shape with an eccentric nucleus bearing indentations. The granules of the corticotroph are about 200 mμ in diameter. They are not as abundant as in the other cell types, but they do occur in association with the Golgi apparatus and near the cell surface. Some of the granules appear as vesicles with a dense core. The Golgi complex (G) of the corticotroph is extensive. Small stacks of cisternae and small vesicles are dispersed through much of the cytoplasm. There are free ribosomes and some isolated cisternae of rough-surfaced endoplasmic reticulum. Mitochondria are spherical to elongated in shape, and a few lysosomes may be found. ACTH is a polypeptide containing 39 amino acids; it has a molecular weight of 4500. It affects primarily the zona fasciculata and zona reticularis of the adrenal cortex, stimulating the cells to produce glucocorticoids.

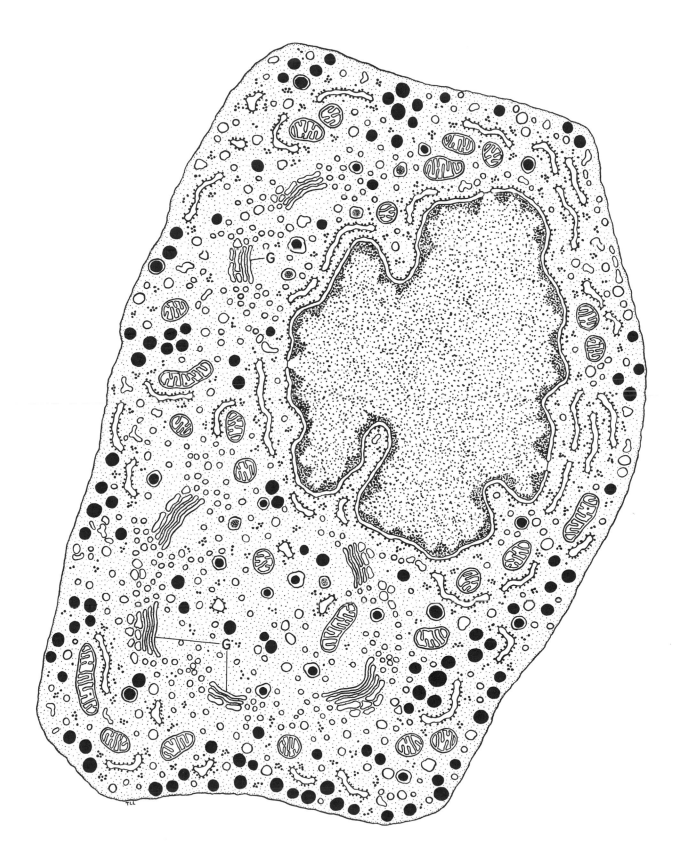

139—PITUITARY: MELANOCYTE-STIMULATING HORMONE (MSH) CELL

The pars intermedia of the pituitary is composed of a few layers of basophilic cells lying between the pars distalis and pars nervosa. In the human, the cells may extend into the neural lobe. The cells of the pars intermedia are polygonal in shape with round eccentric nuclei. A well-developed Golgi apparatus is found near the nucleus. The dense secretory granules are found in the Golgi region and dispersed in the cytoplasm. The cytoplasm also contains a large number of vesicles with clear contents. Mitochondria are distributed around the Golgi apparatus and throughout the cell. The endoplasmic reticulum consists of rough-surfaced cisternae that tend to be more elongated and frequent toward the periphery of the cell. These cells are thought to secrete melanocyte-stimulating hormone (MSH). MSH is a polypeptide similar in chemical structure to ACTH, and it is interesting to note that the cells of the pars intermedia are morphologically similar to the corticotrophs of the anterior pituitary. MSH appears to affect melanocytes, stimulating melanin production which produces darkening of the skin. In amphibians, MSH also causes dispersion of the melanin granules in pigment cells.

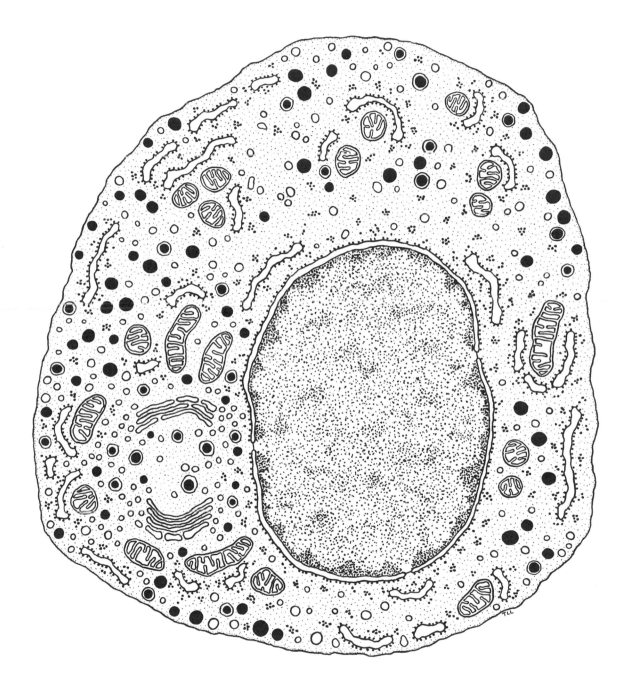

140—PITUITARY: NEUROSECRETORY CELL

The axons in the pars nervosa of the pituitary originate from neurosecretory cells in the hypothalamus, which are located mainly in the supraoptic and paraventricular nuclei. Neurosecretory cells are neurons that are specialized for the secretion of hormonal substances. It is not surprising, therefore, that they have the structural characteristics of most other neurons and in addition contain secretory granules. A well-developed Golgi apparatus occurs near the nucleus. The dense, membrane-bounded neurosecretory granules (NsG), 1200 to 2000 Å in diameter, are most common in the vicinity of the Golgi apparatus and presumably originate from this site. The endoplasmic reticulum is abundant in neurosecretory cells. It consists of short, rough-surfaced cisternae often aggregated to form Nissl bodies. Free ribosomes in clusters are numerous in this region. Small mitochondria are dispersed throughout the cytoplasm. Lysosomes (Ly) and lipofuscin pigment bodies (LPG) are easily distinguished from the smaller neurosecretory granules.

The neural lobe of the pituitary contains two hormones, vasopressin, (antidiuretic hormone, ADH) and oxytocin. These hormones are polypeptides and are believed to be synthesized in the neurosecretory cell bodies. They are segregated in the neurosecretory granules by the Golgi apparatus. The granules are transmitted down the axons and are stored in the axon terminations in the pars nervosa (see Plate 141). The hormones are released from the axon terminals to enter the extensive capillary plexus. The secretory neurons of the hypothalamus, the tract formed by the axons, and the neural lobe of the pituitary constitute the hypothalamo-hypophyseal system.

Other neurosecretory cells are located in the hypothalamus and terminate on vessels of the hypophyseal portal venous system in the median eminence. These cells elaborate releasing factors that are liberated into the capillaries and carried down the stalk to the anterior pituitary to affect the gland cells. The releasing factors were originally thought to be small polypeptides, but more recent evidence indicates they may be polyamines. Five of them cause release of a specific hormone, and one, that for prolactin, inhibits release.

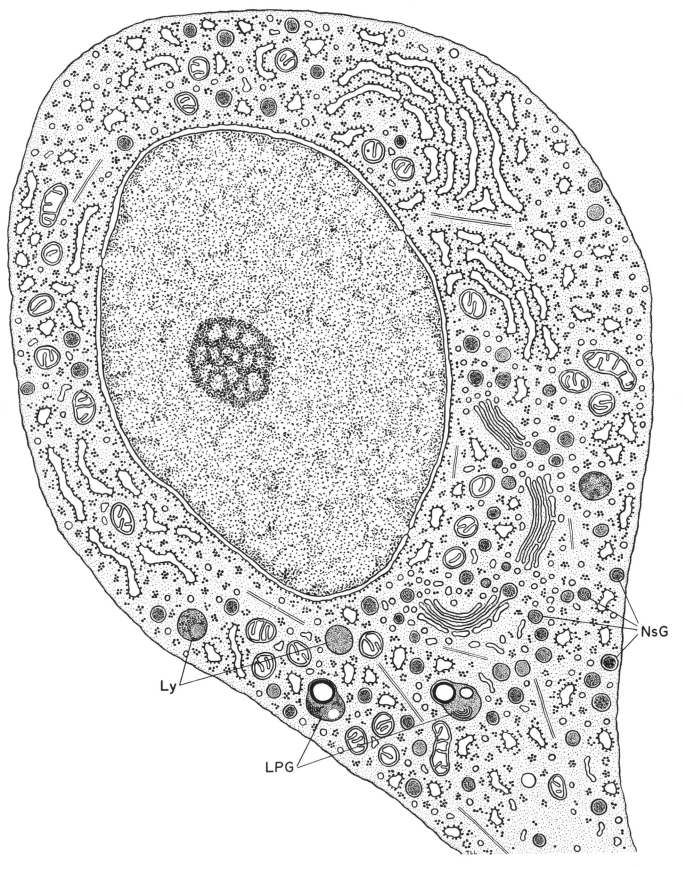

NsG

Ly

LPG

ENDOCRINE SYSTEM

141—PITUITARY: PITUICYTE

The neural lobe of the pituitary consists of the terminal portions of the axons of the neurosecretory cells (Plate 140) of the hypothalamus and the pituicytes. The pituicytes, surrounded by the nerve fibers, are highly irregular in shape with long cytoplasmic processes. The nerve fibers often appear partially embedded in the pituicytes. The relationship of the pituicytes to the axons is similar to that of neuroglia in other parts of the nervous system. An indented nucleus lies in the central region of the cell. A small Golgi apparatus may be found near the nucleus. The endoplasmic reticulum is sparse, consisting of a few short rough-surfaced cisternae scattered in the cytoplasm. Free ribosomes occur in small clusters. Mitochondria are oval to elongated in shape. Filaments are abundant in the cytoplasm and extend into the processes.

As stated previously, the fibers of the hypothalamo-hypophyseal tract that originate from the neurosecretory cells of the hypothalamus terminate in the neural lobe. The axons contain the hormones vasopressin and oxytocin. It has been suggested that each hormone is contained in a separate type of neurosecretory granule (NsG), and variations in the density of the granules are apparent. Some of the large vesicles have a clear content during hormone discharge. Besides the large neurosecretory granules, many small vesicles ($\sim$500 Å in diameter) are found. These structures have been interpreted as synaptic vesicles (SV) containing acetylcholine or, alternatively, as being derived from the large hormone-depleted neurosecretory vesicles.

The appearance of the axon depends on the level at which it is sectioned. Vesicles and granules are most abundant in swellings of the nerve fiber that occur terminally or along its course. Extremely large swellings (upper right of Plate) are known as Herring bodies. Other regions of the axons are devoid of granules but contain microtubules (Mt) or neurotubules, and channels of smooth endoplasmic reticulum. Mitochondria also occur in the axons. Other nerve fibers form synapses on the neurosecretory cell axons (lower left of Plate).

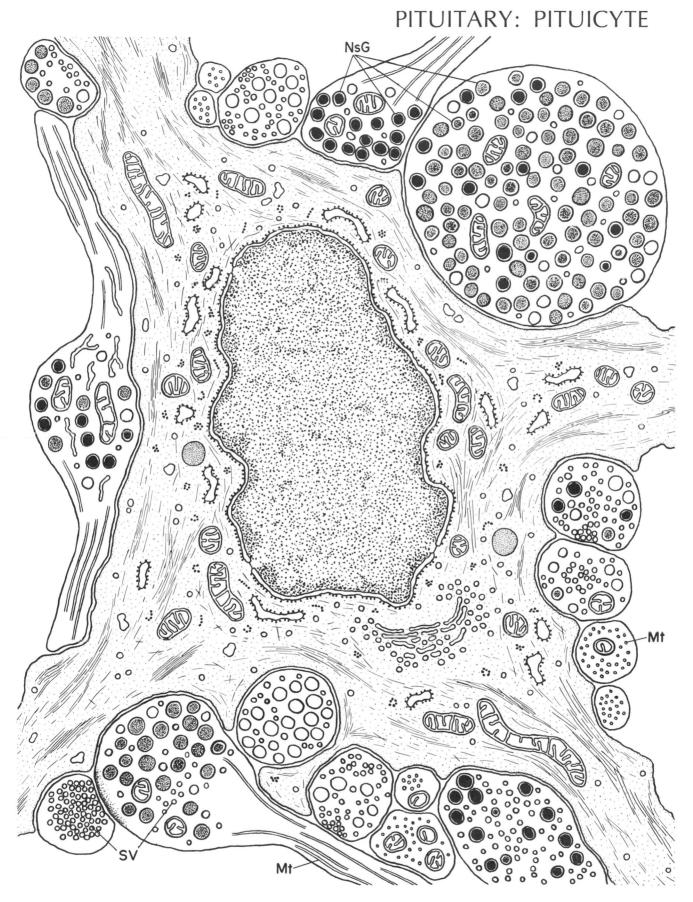

NsG

Mt

SV

Mt

142 — THYROID: FOLLICULAR EPITHELIAL CELL

The follicles of the thyroid gland are lined by a single layer of epithelial cells. The cells are cuboidal with a central nucleus. The apical surface bordering the colloid bears microvilli, whereas the basal surface is flat. The cells have a well-developed, rough-surfaced endoplasmic reticulum. The cisternae are dilated and contain flocculent material. A large Golgi apparatus occurs in a supranuclear position and is composed of a stack of membranous lamellae and many small vesicles. Mitochondria are dispersed throughout the cytoplasm and some free ribosomes are found. The secretory pathway of the glycoprotein thyroglobulin is thought to be the following: synthesis of protein in the endoplasmic reticulum, addition of the polysaccharide moiety in the Golgi apparatus, and release from the apical surface of the cell. The cells also concentrate iodide, which then is oxidized. Tyrosine is iodinated to form mono- and diiodotyrosine which are coupled to form the hormones thyroxin and triiodothyronine. The hormones are stored in the colloid bound to thyroglobulin.

Thyroxin release occurs upon resorption and degradation of follicular colloid. Colloid is engulfed by thin pseudopods that extend into the follicular lumen. Newly formed colloid droplets (CD) of low density occur in the apical cytoplasm. Membrane-bounded dense granules (Gr) of different sizes are found in the same area and in association with the Golgi apparatus. Some of the dense granules fuse with the colloid droplets, which become progressively denser and smaller toward the base of the cell. The dense bodies have been shown to contain acid phosphatase activity. Acid phosphatase activity appears within the colloid droplets when the dense bodies fuse with them. Dense structures rich in acid phosphatase activity and possibly representing residual bodies are found in the base of the cell. The addition of the hydrolases contained in the dense bodies to the colloid droplets could produce degradation of thyroglobulin and release of hormones.

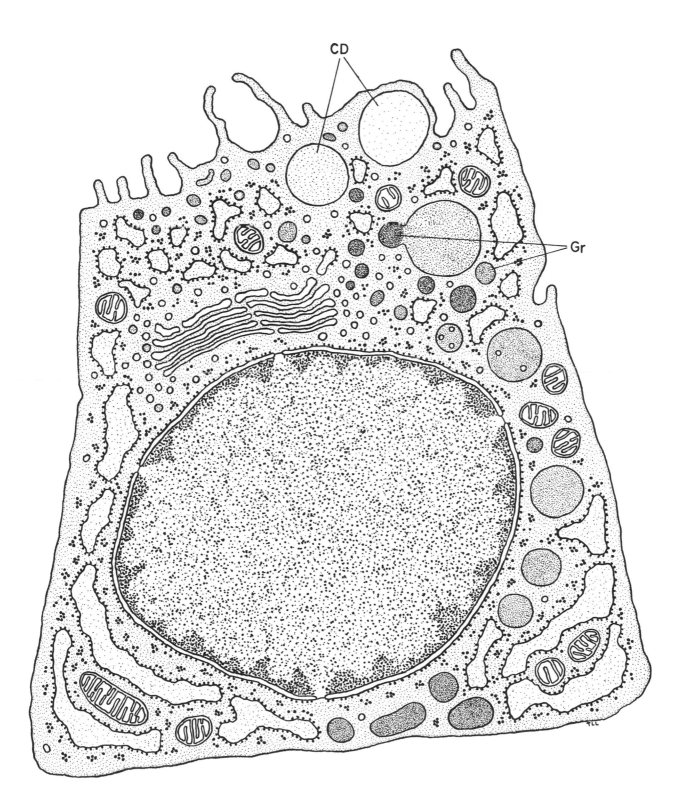

143 — THYROID: PARAFOLLICULAR CELL

Parafollicular or light cells are located either singly or in groups at the periphery of the follicle. They are separated from the lumen by the follicular cells but are included within the basement membrane of the follicle. The cells are round, oval, or polyhedral in shape and have a centrally located nucleus. They have an extensive Golgi apparatus situated adjacent to the nucleus. Small vesicles and dense, membrane-bounded granules are found in association with the Golgi complex. The granules are 1000 to 2000 Å in diameter. They are distributed throughout the cytoplasm and are sometimes concentrated toward the surface of the cell facing the interfollicular space. The content of the granules varies in density from moderate to opaque and is often separated from the limiting membrane by a clear halo. Elongated cisternae of rough-surfaced endoplasmic reticulum are present, especially below the cell surface farthest away from the interfollicular space. Free ribosomes are also common. Mitochondria with a dense matrix and large dense bodies resembling lysosomes are found in the cytoplasm. It is now known that the hormone calcitonin or thyrocalcitonin is produced by the thyroid. The parafollicular cells are prime candidates for the production of calcitonin in view of their obvious secretory activity. It has also been observed that parafollicular cells are engaged in the metabolism of biogenic amines, and in this regard they have similarities to the chromaffin cells of the adrenal medulla (Plates 149, 150).

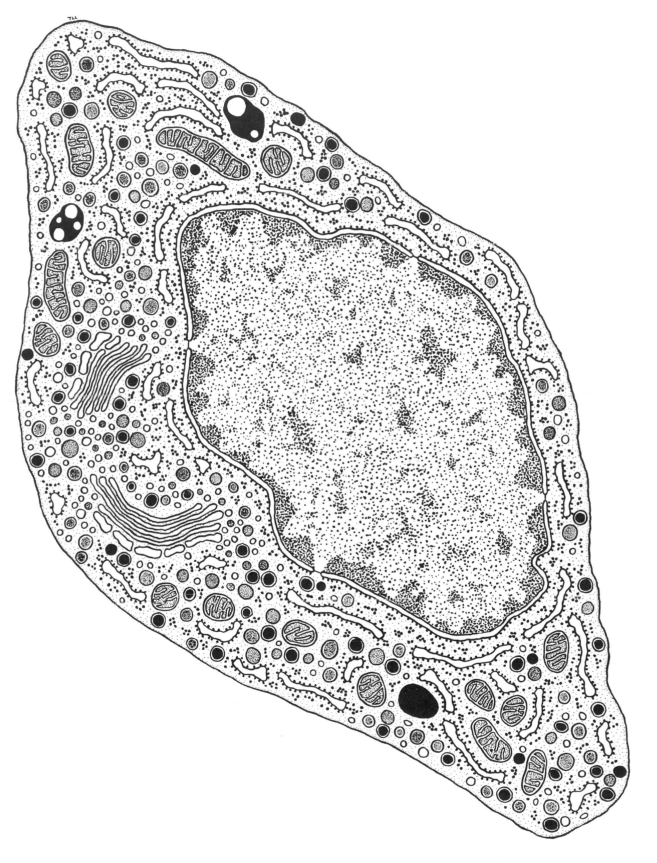

144—PARATHYROID: PRINCIPAL CELL

There are usually four small parathyroid glands embedded in the capsule of the thyroid. The glands are composed of densely packed cells of which there are two types: principal cells and oxyphilic cells. The principal or chief cells of the parathyroid glands are polygonal in shape with a central nucleus. A prominent Golgi apparatus is situated adjacent to the nucleus. Vesicles and small dense granules are associated with the Golgi complex. Larger granules (Gr) are found elsewhere in the cytoplasm, often concentrated near the plasma membrane. These granules are extremely dense and may be spherical, oval, or dumbbell in shape. The granules are thought to correspond to the bodies that can be observed with the light microscope when chrome alum-hematoxylin, aldehyde fuchsin, and silver stains are used. They are believed to be secretory granules containing parathyroid hormone, which is a polypeptide with a molecular weight of 8500. Cisternae of rough-surfaced endoplasmic reticulum are found, sometimes occurring in prominent stacks. Free ribosomes are also present. Mitochondria are small, oval, or round in profile and are numerous. Lipofuscin pigment bodies are seen. Masses of glycogen granules (Gly) occur and are especially large in resting or inactive chief cells. The latter cells have a small Golgi apparatus and few secretory granules.

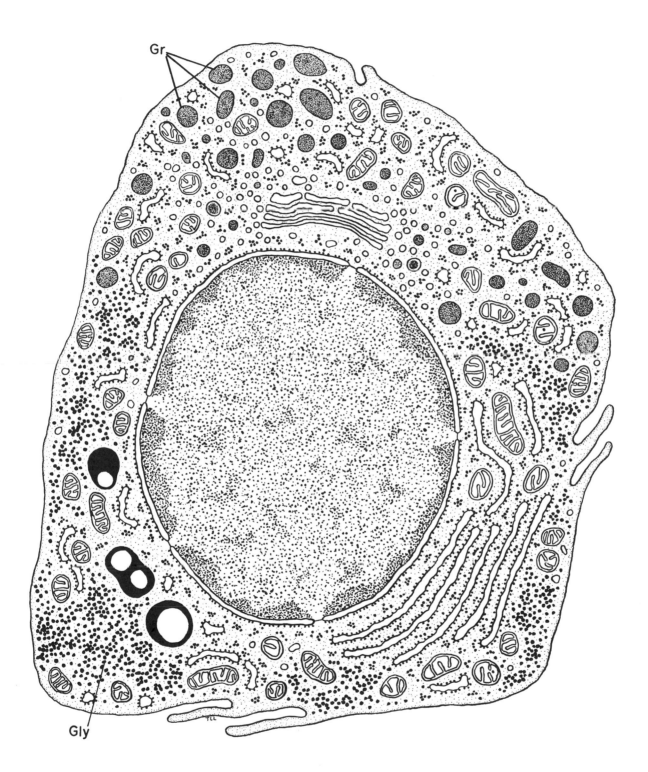

145—PARATHYROID: OXYPHIL CELL

Oxyphil cells are larger than, but not nearly as numerous as, chief cells. They occur singly or in small clumps and are polygonal in shape with a central nucleus. The cells are unusual in that the cytoplasm in largely occupied by mitochondria. Large amounts of oxidative enzyme activity have been detected histochemically in these cells. The mitochondria are round to elongated in profile with closely spaced cristae. Glycogen granules are found in the narrow cytoplasmic spaces between the closely packed mitochondria. Other organelles are sparse or absent. A few rough or smooth membranous profiles can be found. Oxyphil cells seem to increase in number with increasing age, but their function is unknown. They contain few or no secretory granules and are probably not engaged in the production of parathyroid hormone. Cells with structural characteristics intermediate between chief and oxyphil cells have also been observed in the parathyroid gland.

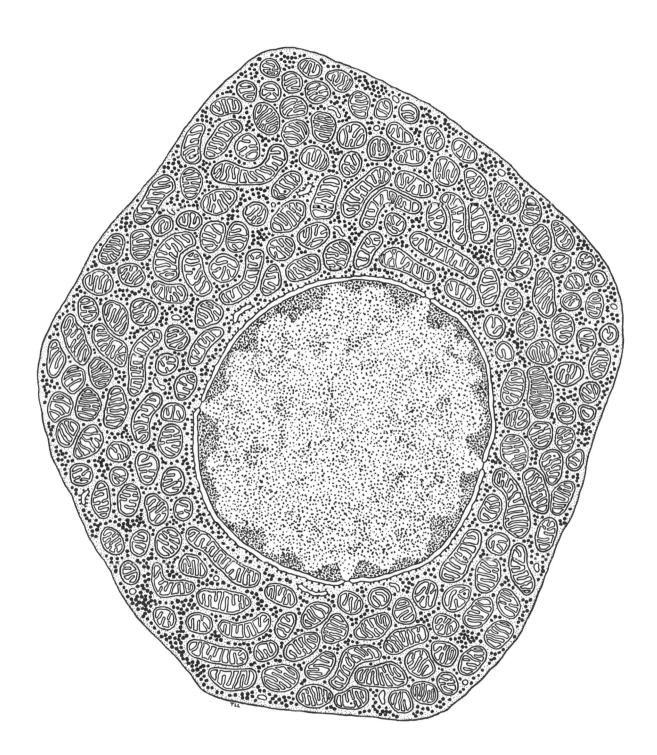

ENDOCRINE SYSTEM

146—ADRENAL: CELL OF THE ZONA GLOMERULOSA

The adrenal cortex is composed of three distinct zones, the zona glomerulosa, zona fasciculata, and zona reticularis. The zona glomerulosa secretes the mineralocorticoids, deoxycorticosterone and aldosterone, which are involved in mineral metabolism. The zona fasciculata and zona reticularis secrete the glucocorticoids, cortisone and cortisol or corticosterone, which are concerned with regulation of carbohydrate, protein, and fat metabolism.

The cells of the zona glomerulosa are columnar and have a spherical nucleus with a well-developed nucleolus. The cell contour is smooth except adjacent to the subendothelial space and to some dilated intercellular spaces. In these regions, the plasma membrane is thrown up into folds and microvilli and has coated pinocytotic vesicles associated with it. A basement lamina (BL) coats the perivascular surface. A prominent feature of the cell is the extensive smooth-surfaced endoplasmic reticulum (SER) which occurs as a network of branching and anastomosing tubules. Some of the tubules closely parallel the surfaces of mitochondria and lipid droplets. There are a few short segments of rough-surfaced endoplasmic reticulum and also some free cytoplasmic ribosomes which may occur in clusters, spirals, or linear chains. Mitochondria (M) are round, oval, or elongated with cristae in the form of lamellar infoldings of the inner mitochondrial membrane. Lipid droplets (LD) are spherical or irregular in shape. The lipid droplets are shown here as opaque, but they may be dissolved out by routine fixation and embedding, leaving clear spaces. As mentioned previously, mitochondria and smooth-surfaced cisternae sometimes occur in close relationship to the lipid droplets. A well-developed Golgi apparatus is present adjacent to the nucleus. Other structures include lysosomes, multivesicular bodies, and glycogen granules.

The subcellular localization of the enzymes participating in aldosterone synthesis have been determined by cell fractionation procedures. Synthesis of cholesterol from acetate takes place in the smooth endoplasmic reticulum, while conversion of cholesterol to pregnenolone takes place in the mitochondria. The enzymes associated with the synthetic pathway from pregnenolone to progesterone to deoxycorticosterone are associated with the smooth endoplasmic reticulum, while those enzymes converting deoxycorticosterone to corticosterone to 18-hydroxycorticosterone to aldosterone are again located in the mitochondria.

Aldosterone results in excretion of potassium by the kidney and retention of sodium and water. Thus, a sodium-deficient diet stimulates heightened secretory activity by the zona glomerulosa. After this diet, there is a marked increase in the amount of smooth-surfaced endoplasmic reticulum. An increase in the amount of this organelle could be expected to facilitate aldosterone synthesis by increasing the activity or amounts of the enzymes involved in the intermediate steps of aldosterone synthesis, and of those enzymes involved in cholesterol synthesis. Rough-surfaced endoplasmic reticulum is also increased in stimulated animals, and this could be related to the elaboration of more smooth membranes. Mitochondria do not undergo marked changes, but some acquire a dense mass within the matrix, which could represent lipid or intermediates in steroid biosynthesis.

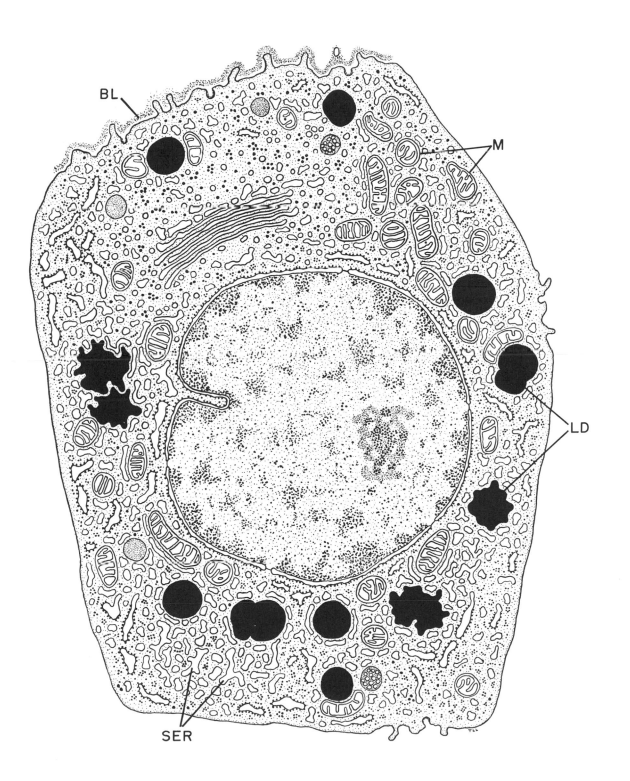

147—ADRENAL: CELL OF THE ZONA FASCICULATA

The cells of the zona fasciculata are large polyhedral cells with a central nucleus. The nucleus contains peripherally condensed chromatin, a nucleolus, and an internal fibrous lamina (FL). The branching and anastomosing tubules of smooth-surfaced endoplasmic reticulum (SER) are extensive. The rough-surfaced endoplasmic reticulum (ER) is more abundant in this zone than in the zona glomerulosa, consisting of lamellar stacks of ribosome-studded cisternae. Clusters of ribosomes also occur free in the cytoplasm. The rough-surfaced cisternae are continuous with the smooth tubular system. The mitochondria (M) are large and usually spherical in shape. The cristae occur as vesicular inpocketings of the inner mitochondrial membrane and as vesicles apparently free in the mitochondrial matrix. The vesicular cristae of the zona fasciculata contrast with the lamellar cristae in the zona glomerulosa. Lipid droplets (LD) are common and are larger than in the zona glomerulosa. The relationship between lipid droplets and mitochondria and smooth endoplasmic reticulum is again observed in the zona fasciculata. These cells have a well-developed Golgi apparatus, lysosomes (Ly), and lipofuscin pigment granules (LPG). The latter are membrane-bounded structures with a moderately dense matrix within which are embedded dense granules and larger clear globules of lipid. Glycogen is absent in the zona fasciculata. A pair of centrioles is sometimes seen adjacent to the nucleus and near the Golgi apparatus. Microvilli extend into the subendothelial space and a few pinocytotic invaginations occur at the cell surface in this region. A basement lamina covers the subendothelial surface.

By means of biochemical fractionation studies, the various products and enzymes of steroid biosynthesis have been localized within subcellular compartments. Cholesterol is absorbed from the blood, or synthesized in the adrenal from acetate, and stored in the lipid droplets. The first step in steroid synthesis, the conversion of cholesterol to pregnenolone, a desmolase reaction, takes place in the mitochondria. The enzyme responsible for the conversion of pregnenolone to progesterone (3β-ol-dehydrogenase isomerase) is found in the microsomal fraction (smooth endoplasmic reticulum). 17-Hydroxylase, which converts progesterone to 17α-hydroxyprogesterone, and 21-hydroxylase, which converts 17α-hydroxyprogesterone to 11-deoxycortisol and progesterone to 11-deoxycorticosterone (DOC), also occur in the smooth endoplasmic reticulum. The 11-hydroxylase responsible for the final steps of conversion of DOC to corticosterone, and of 11-deoxycortisol to cortisol, is localized in the mitochondria.

ACTH, which initiates synthesis and subsequent release of steroid hormones, has a profound effect on the structure of the cells of the zona fasciculata, although the changes differ somewhat from species to species. In general, administration of ACTH causes depletion of lipid material from the adrenal, presumably because of the utilization of stored hormone precursors. Changes in mitochondria include mainly an increase in the number of vesicular cristae, providing a greater surface area available for enzymatic activity. The appearance of dense masses in mitochondria and an increase in density of the matrix have also been observed. Similarly, the smooth endoplasmic reticulum proliferates, thereby enhancing the intermediate steps in steroid biosynthesis, or facilitating cholesterol biosynthesis.

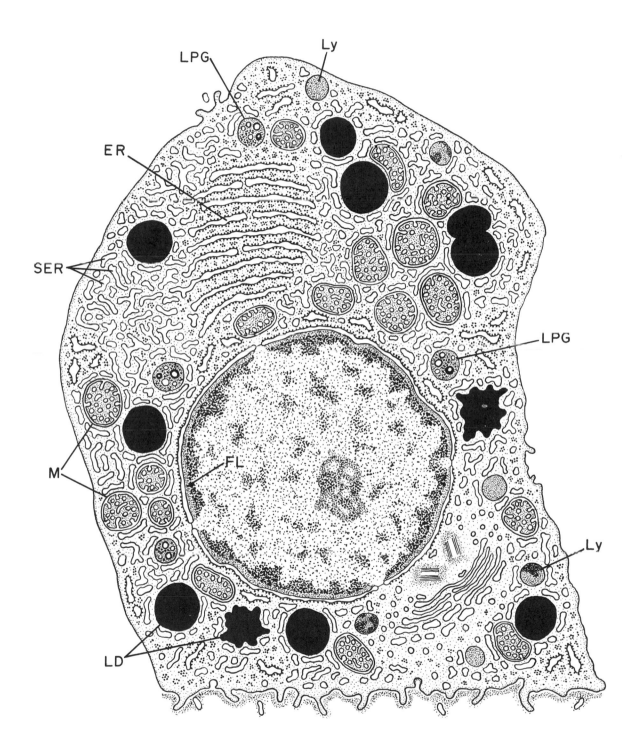

148—ADRENAL: CELL OF THE ZONA RETICULARIS

The cells of the zona reticularis have many of the features observed in the zona fasciculata, including tubules of smooth-surfaced endoplasmic reticulum, stacks or whorls of rough-surfaced lamellae, cytoplasmic ribosomes, Golgi apparatus, lipid droplets, and lysosomes. They differ in the structure of the mitochondria and in the number and size of lipofuscin pigment bodies. The mitochondria are more often elongated in this zone of the adrenal. Two types of cristae are present: vesicular cristae and lamellar invaginations of the inner mitochondrial membrane. Glycogen, seen also in the zona glomerulosa, is present in the zona reticularis.

The lipofuscin pigment granules (LPG) are present in large numbers and may attain considerable size. They are irregular in shape and consist of a granular matrix within which are embedded droplets of varying size. These droplets, probably representing lipid, may be lucent or may have a dense rim. When they occur at the edge of the pigment body, they produce a bulge in its contour. Smaller vesicles and vacuoles containing material similar to the granular component of the lipofuscin body occur in the cytoplasm, sometimes immediately adjacent to the large body. The pigment contained in the granules is responsible for the golden brown color of these adrenal cells when viewed with the light microscope.

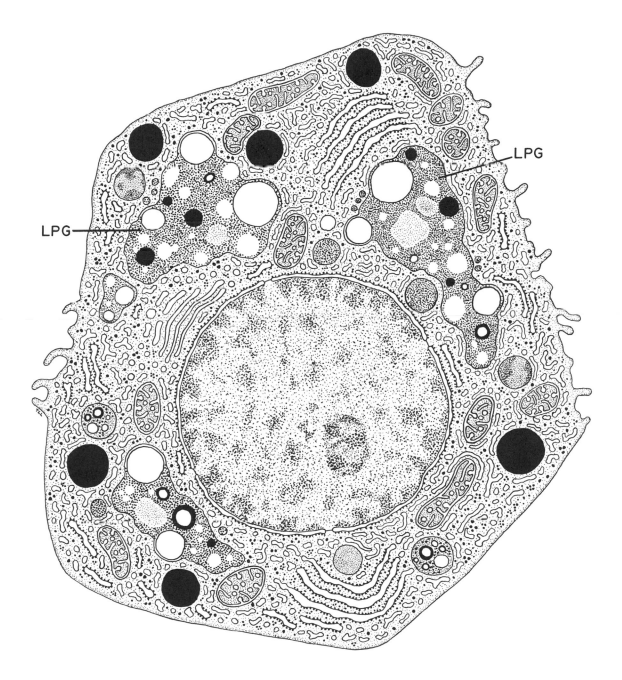

LPG

LPG

149—CHROMAFFIN CELL—NOREPINEPHRINE-STORING

The adrenal medulla is composed of chromaffin cells and a few sympathetic ganglion cells. There are two types of chromaffin cells: those containing norepinephrine and those storing epinephrine. The cells of the adrenal medulla arise from neural crest tissue and are innervated by preganglionic sympathetic fibers which control the secretion of catecholamines into the blood stream.

Chromaffin cells are polyhedral in shape with a large rounded nucleus. One or more nucleoli are present. The cytoplasm contains a large population of membrane-bounded granules 1000 to 3000 Å in diameter. The two types of chromaffin cells are distinguished on the basis of differences in the granules. In tissue fixed in glutaraldehyde and osmium, the granules of norepinephrine-containing cells are opaque. The membrane enclosing the granule is sometimes dilated and the granule is located eccentrically in the membrane sac.

Microvilli are found on some regions of the cell surface. A large Golgi complex occurs near the nucleus. Small, immature granules as well as a pair of centrioles are found in this region. Mitochondria, lysosomes, multivesicular bodies, vacuoles, and cisternae of rough-surfaced endoplasmic reticulum are distributed in the cytoplasm among the granules. Most of the cisternae of endoplasmic reticulum are short, but a stack of more elongated flattened cisternae may occur.

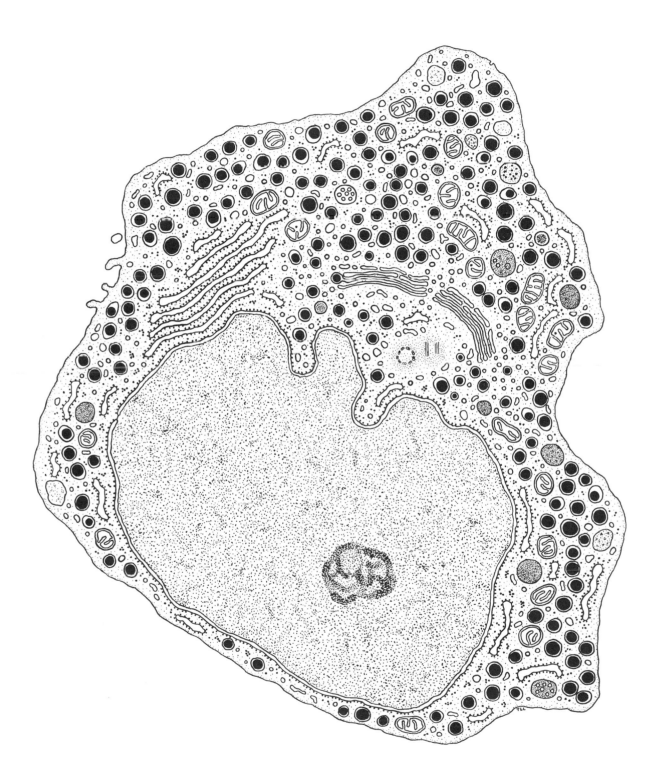

150—CHROMAFFIN CELL—EPINEPHRINE-STORING

Chromaffin cells containing epinephrine are similar in general structure to norepinephrine cells. They contain the same complement of organelles and differ only in the structure of the granules. The granules of the epinephrine-containing cell are moderately dense, but not opaque, and finely granular in texture. A thin symmetrical halo occurs between the granule and the enclosing membrane.

Chromaffin cells similar to those in the adrenal medulla also occur in widely scattered accumulations that constitute the paraganglia or chromaffin system. Groups of cells are found in the retroperitoneum, the largest being the para-aortic bodies of Zuckerkandl. Chromaffin cells are also found in the kidney, ovary, testis, liver, heart, and gastrointestinal tract. The carotid and aortic bodies, which function as chemoreceptors, contain similar types of cells.

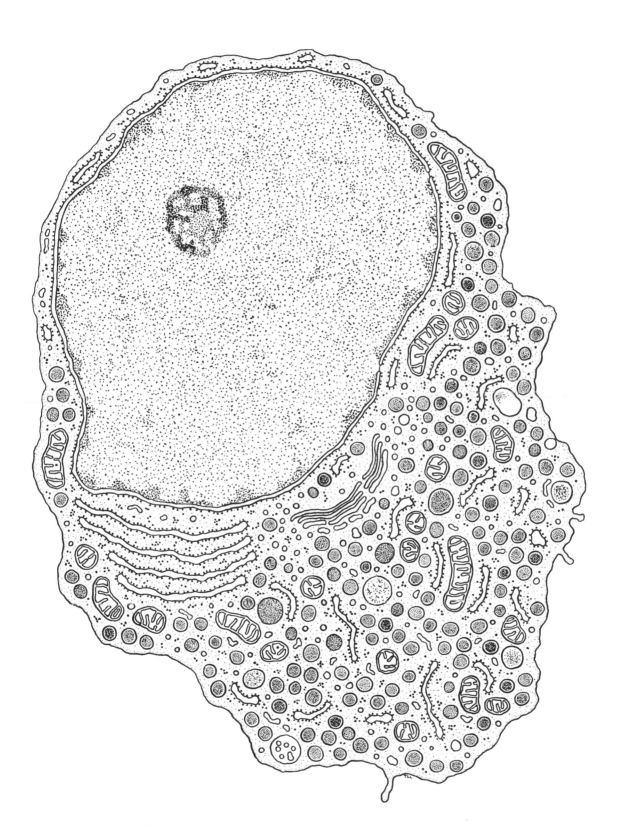

151—ISLET OF LANGERHANS: ALPHA CELL

The islets of Langerhans are small groups of endocrine cells scattered throughout the pancreas. Four cell types can be distinguished in the islets on the basis of cytological structure.

Alpha cells are polygonal in shape and filled with membrane-bounded secretory granules. The granules consist of a central, spherical core that is electron-opaque and an outer halo of low density between the core and membrane. With aldehyde fixation, the halo is often of medium density. The granules of alpha cells are slightly larger than those of beta cells. The alpha cells are thought to secrete the hormone glucagon, a polypeptide that has a glycogenolytic effect and raises blood glucose.

The spherical nucleus is sometimes indented and contains a nucleolus. Near the nucleus is a well-developed Golgi apparatus. Some of the Golgi vesicles contain dense material and may represent precursors of the secretory granules. Free ribosomes and short cisternae of rough-surfaced endoplasmic reticulum are found in the cytoplasm. Mitochondria are usually oval or elongated in profile. A few lysosomes and lipofuscin pigment bodies occur in alpha cells.

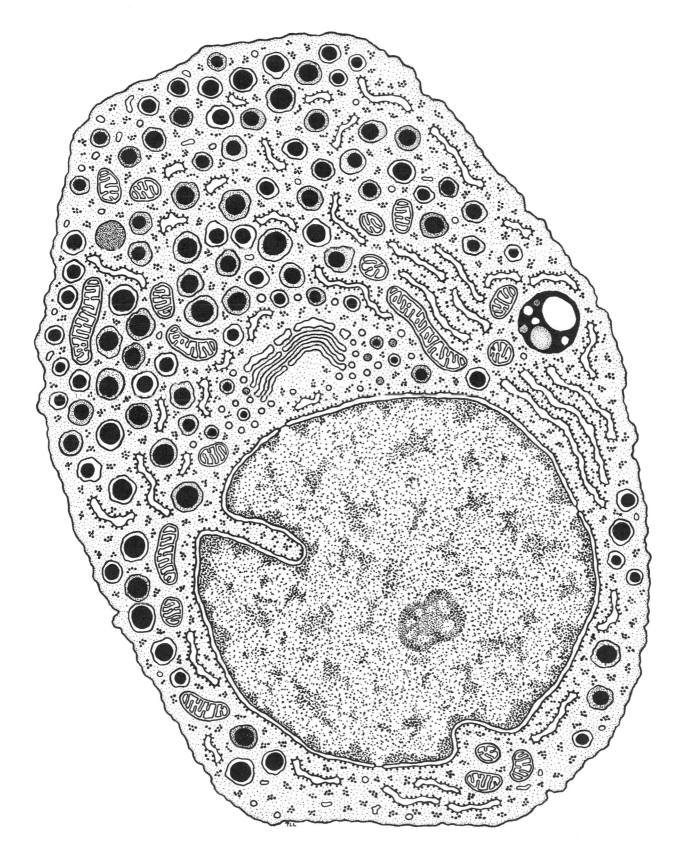

152—ISLET OF LANGERHANS: BETA CELL

Beta cells are more numerous than alpha cells and tend to be concentrated in the center of the islet. The beta cell secretory granules (SG) are contained in dilated membranous sacs. In some species the granules are round, but in others, including man, they appear to be composed of one or more rectangular to polygonal crystals. The granules are extremely dense and at high magnification they have a periodic internal structure. The region between the dense crystal and the limiting membrane is of low density.

The beta cells have a round central nucleus that is relatively smooth in contour. The Golgi apparatus is in a juxtanuclear position and is larger than that of the alpha cell. Dense amorphous material is seen in some of the Golgi vesicles. Mitochondria are larger and more numerous than those of the alpha cell but the endoplasmic reticulum and ribosomes are not as prominent. The endoplasmic reticulum is composed of short cisternae or vesicular profiles except in cells with few granules, where the cisternae are elongated and occur in stacks. Lysosomes and lipofuscin pigment granules are relatively abundant in beta cells. The beta cells are the source of the hormone insulin, a small protein molecule that lowers blood glucose and accelerates glycogenesis.

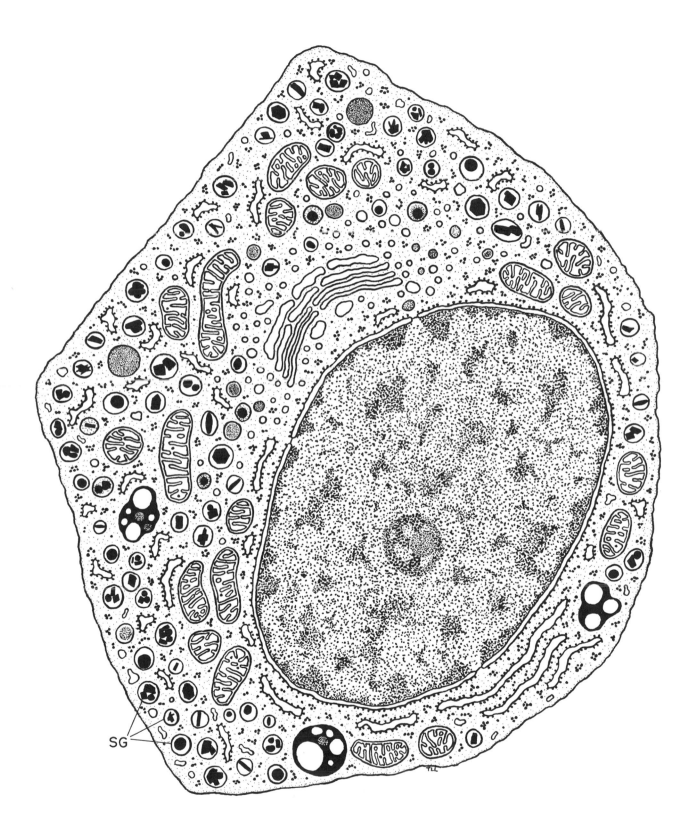

SG

153—ISLET OF LANGERHANS: DELTA CELL

Although there is no physiological evidence for a third hormone-secreting cell type in the pancreas, the delta cell can be distinguished morphologically from alpha and beta cells. It is less abundant than the alpha and beta cells and comprises only about 5 per cent of the cells. It has been suggested that delta cells are altered alpha cells, and transitional stages have been described (Like).

Except for the granules, the delta cells are similar in structure to the alpha cells. The overall dimensions of the granules of the two cell types are the same. The delta cell granule, however, lacks a dense core. Instead, the granular content fills the vesicle and varies in density from light and flocculent to fairly dense and compact. The cells are polygonal in shape with a round, central nucleus. Besides the granules, the cytoplasm contains short cisternae of endoplasmic reticulum, mitochondria, and ribosomes. A Golgi apparatus near the nucleus has small granules associated with it.

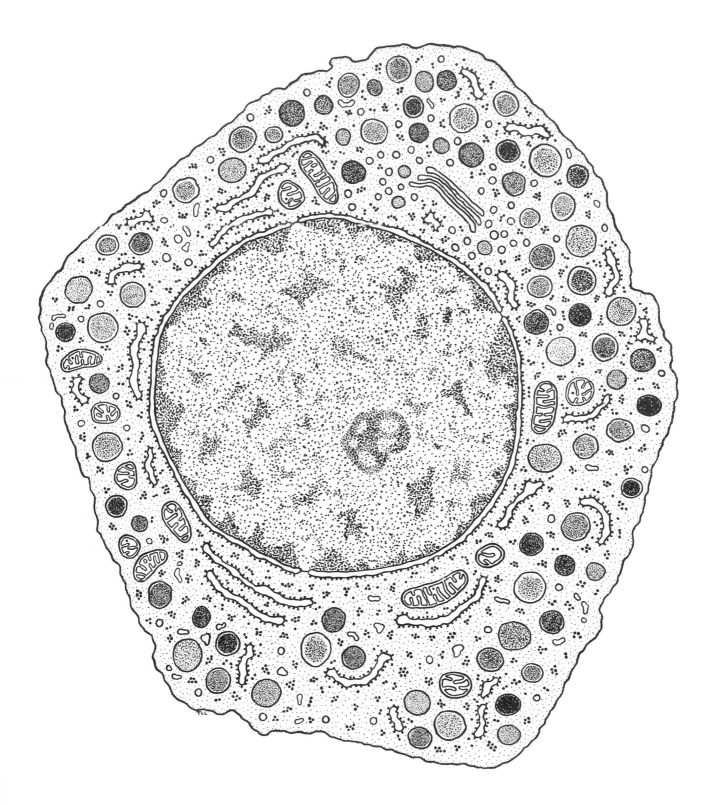

154—ISLET OF LANGERHANS: CHROMOPHOBIC (C) CELL

The fourth morphologically distinct cell type found in the islet is known as the C cell or chromophobic cell. It differs from the others in that it lacks granules, or has only a very few in the cytoplasm. Because of the absence of granules, the cell does not stain with special methods as do the granular cell types; hence the name chromophobic. On the other hand, large vesicles with clear contents are numerous. These structures may be elements of a smooth-surfaced endoplasmic reticulum. The nucleus is centrally situated in the polygonal or angular cells. Elements of rough-surfaced endoplasmic reticulum are few, although there are a number of free ribosomes. A small Golgi apparatus and an occasional pigment body are found in the cytoplasm. The C cells are relatively unspecialized structurally and may represent the end stage in the transition of alpha cells (Like).

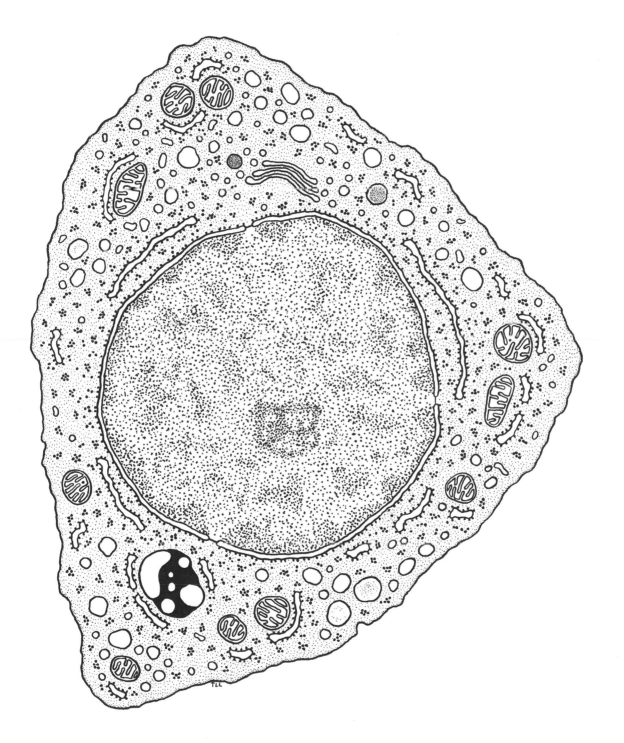

155—PINEAL: PINEALOCYTE

The pineal is a small body connected by a short stalk to the caudal end of the diencephalic roof. It is composed of two cell types: the pinealocytes (parenchymal cells, chief cells) and the interstitial cells.

Pinealocytes are stellate cells arranged in cords and clusters with cytoplasmic processes that end in bulbous swellings (lower right of Plate). The cells have a central nucleus with a nucleolus. Two conspicuous features are seen in the cytoplasm. The first is the presence of large numbers of microtubules (Mt). In the perikaryon, they extend in all directions between organelles, while in the processes they are parallel to the long axis. Secondly, the cells have an extensive smooth-surfaced endoplasmic reticulum (SER). Short tubular and vesicular cisternae fill much of the cytoplasm. Only a small number of short cisternae of rough-surfaced endoplasmic reticulum and polyribosomes are present. A large Golgi apparatus occurs adjacent to the nucleus, and a pair of centrioles is usually found nearby. Small vesicles are especially abundant in the Golgi region. In addition, some small, dense, membrane-bounded granules (Gr) are found in this region as well as throughout the cell and its processes. Mitochondria are common and relatively large with transverse cristae oriented across the width of the organelles. Lysosomes and lipid droplets are also seen in the cytoplasm.

Melatonin, a 5-hydroxyindole synthesized from 5-hydroxytryptamine, and an enzyme necessary for the synthesis of melatonin, 5-hydroxyindole-O-methyltransferase (HIOMT), have been identified in the pineal. In amphibians, melatonin causes the aggregation of pigment granules in melanocytes, thus producing blanching of the skin, and may be an antagonist of melanocyte-stimulating hormone. Light is thought to have an effect on the pineal via the superior cervical sympathetic ganglia and the sympathetic nerve fibers terminating in the pineal. Exposure to light decreases the HIOMT activity and the synthesis of melatonin. Since melatonin inhibits the ovary, exposure to light stimulates the ovary by inhibiting melatonin secretion. Thus, the pineal might play a role in regulation of some of the rhythmic activities of the endocrine system.

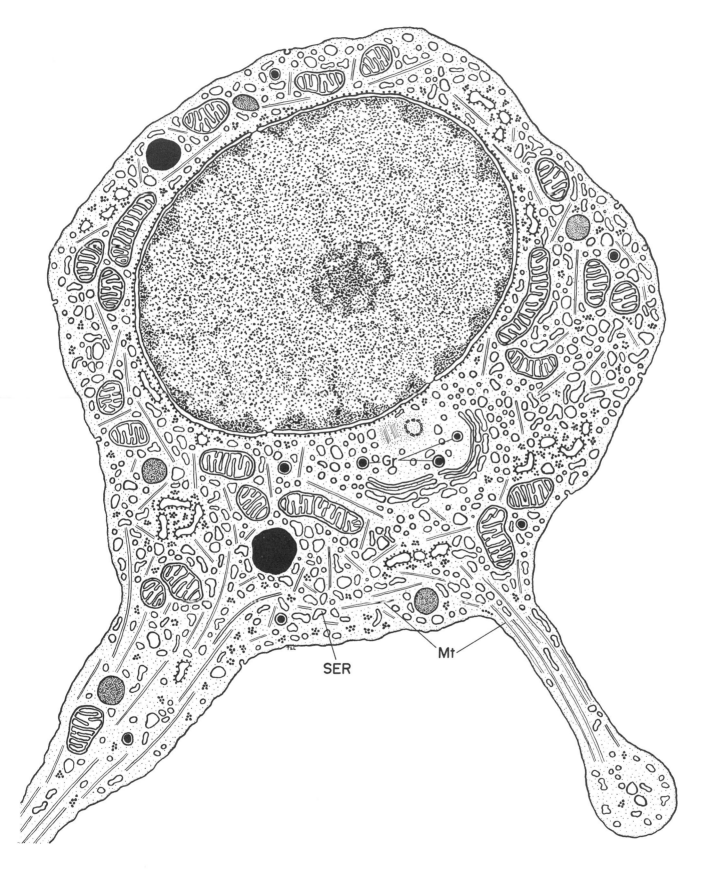

SER

Gr

Mt

156—PINEAL: INTERSTITIAL CELL

The interstitial cells are scattered among the pinealocytes and, like them, are highly irregular in shape. They consist of a cell body with a round or oval nucleus and long branching processes. Cytoplasmic filaments (Fl) occur in bundles that course into the cytoplasmic processes. A number of short cisternae of rough-surfaced endoplasmic reticulum are found, and clusters of ribosomes are abundant free in the hyaloplasm. Larger glycogen granules (Gly) are also common, especially in large masses within the processes. The Golgi complex is not extensive, but multivesicular bodies and lysosomes (Ly) are frequently encountered. A pair of centrioles or even a cilium may occur near the Golgi complex. Small mitochondria are distributed throughout the cell. Interstitial cells bear a resemblance to astrocytes (Plates 161, 162), and some investigators consider them to be a type of glial cell.

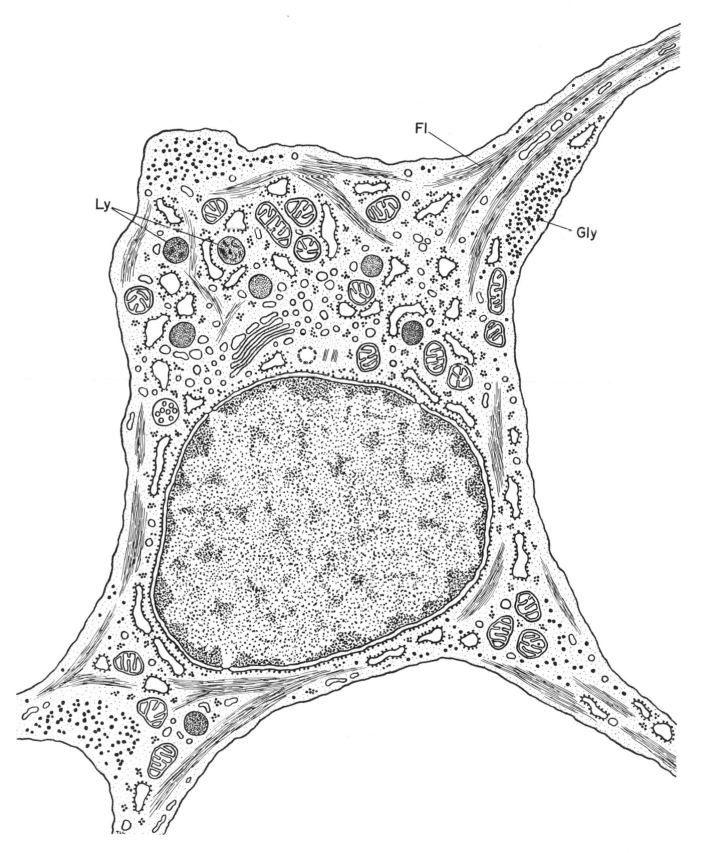

REFERENCES

Al-Lami, F. and R. G. Murray, 1968. Fine structure of the carotid body of *Macaca mulata* monkey. J. Ultrastruct. Res., 24:465–478.

Anderson, E., 1965. The anatomy of bovine and ovine pineals. Light and electron microscopic studies. J. Ultrastruct. Res., Suppl. 8:1–80.

Barer, R. and K. Lederis, 1966. Ultrastructure of the rabbit neurohypophysis with special reference to the release of hormones. Z. Zellforsch., 75:201–239.

Brenner, R. M., 1966. Fine structure of adrenocortical cells in adult male rhesus monkeys. Amer. J. Anat., 119:429–454.

Coupland, R. E., 1965. Electron microscopic observations on the structure of the rat adrenal medulla. I. The ultrastructure and organization of chromaffin cells in the normal adrenal medulla. J. Anat., 99:231–254.

Ekholm, R. and L. E. Ericson, 1968. The ultrastructure of the parafollicular cells of the thyroid gland in the rat. J. Ultrastruct. Res., 23:378–402.

Kobayashi, Y., 1965. Functional morphology of the pars intermedia of the rat hypophysis as revealed with the electron microscope. II. Correlation of the pars intermedia with the hypophyseo-adrenal axis. Z. Zellforsch., 68:155–171.

Kurosumi, K. and Y. Oota, 1968. Electron microscopy of two types of gonadotrophs in the anterior pituitary glands of persistent estrous and diestrous rats. Z. Zellforsch., 85:34–46.

Like, A. A., 1967. The ultrastructure of the secretory cells of the islets of Langerhans in man. Lab. Invest., 16:937–951.

Long, J. A. and A. L. Jones, 1967. The fine structure of the zona glomerulosa and the zona fasciculata of the adrenal cortex of the opossum. Amer. J. Anat., 120:463–488.

Long, J. A. and A. L. Jones, 1967. Observations on the fine structure of the adrenal cortex of man. Lab. Invest., 17:355–370.

Long, J. A. and A. L. Jones, 1970. Alterations in fine structure of the opossum adrenal cortex following sodium deprivation. Anat. Rec., 166:1–26.

Munger, B. L. and S. I. Roth, 1963. The cytology of the normal parathyroid glands of man and Virginia deer. A light and electron microscopic study with morphologic evidence of secretory activity. J. Cell Biol., 16:379–400.

Smith, R. E. and M. G. Farquhar, 1966. Lysosome function in the regulation of the secretory process in cells of the anterior pituitary gland. J. Cell Biol., 31:319–348.

Wetzel, B. K., S. S. Spicer, and S. H. Wollman, 1965. Changes in fine structure and acid phosphatase localization in rat thyroid cells following thyrotropin administration. J. Cell Biol., 25:593–618.

Yamada, K. and K. Yamashita, 1967. An electron microscopic study on the possible site of production of ACTH in the anterior pituitary of mice. Z. Zellforsch., 80:29–43.

Zambrano, D. and E. DeRobertis, 1966. The secretory cycle of supraoptic neurons in the rat. A structural-functional correlation. Z. Zellforsch., 73:414–431.

NERVOUS SYSTEM

NERVOUS SYSTEM

157—NEURON

Neurons are large cells generally consisting of a cell body containing a nucleus, a number of branching dendrites (Den) originating from one pole of the cell, and a single straight axon (Ax) extending from the opposite pole. It may be surprising, in view of the complex functions performed by the nervous system, that neurons from different regions are basically similar in cytologic structure. However, as emphasized by Peters, Palay, and Webster, the functional specificity of the nerve cell is primarily the result of its relationship to other neurons as determined by its position in the nervous system and the anatomical form of its dendritic tree and axon.

Much of the cell body is occupied by the large nucleus (N), which is usually round but may have some shallow indentations. The double-layered nuclear envelope is perforated by many pores bridged by diaphragms. The chromatin material is not highly condensed but occurs in small aggregates, giving a variegated appearance to the nucleus. A large, dense, spherical nucleolus (Nl) consists of filamentous and granular components arranged into coarse, branching and anastomosing strands. In females of some species, a nucleolar satellite (NS) containing the heterochromatic X chromosome is found. The satellite consists of coarse, coiled fibrils and is attached to the fibrous component of the nucleolus.

The rough-surfaced endoplasmic reticulum of neurons is organized into aggregations of cisternae known as Nissl bodies or Nissl substance (NiS). The cisternae are flattened and studded with ribosomes. In some places the cisternae are stacked in parallel, but elsewhere they are more irregularly arranged, or they branch.

Ribosomes (R) are common free in the cytoplasm in the interstices between cisternae, as well as throughout the cell. They are arranged in clusters or rosettes to form polysomes. Short, isolated cisternae of endoplasmic reticulum not organized into Nissl bodies are also found throughout the cytoplasm. Smooth-surfaced endoplasmic reticulum (SER) is fairly extensive and takes the form of tubules or cisternae. Flattened, subsurface cisternae (SsC) are found in some places immediately beneath the plasma membrane. Connections between smooth and rough-surfaced cisternae are common.

The Golgi apparatus (G) consists of a short stack of flattened cisternae with small vesicles (V) ($\sim$500 Å in diameter) clustered around it. Some of the vesicles have striae that look like spines or bristles projecting radially from the outer surface of the vesicle (coated vesicles). Other vesicles are larger (600 to 800 Å) and enclose a dense granule. These dense core vesicles (DV) are thought to contain catecholamines. Golgi complexes are common near the nucleus, but they also occur elsewhere throughout the cytoplasm.

Lysosomes (Ly) are numerous in the perikaryon. Most are 0.2 to 0.5 μ in diameter and contain finely granular, moderately dense material. Others have a heterogeneous content which includes membranes, vacuoles, and dense droplets. Intermediate stages between lysosomes and lipofuscin pigment granules (LPG) are found. The latter are large, lobulated, and contain vacuoles and membranous lamellae in a finely granular matrix. Multivesicular bodies also occur in neurons.

Mitochondria are small, and round to elongated in profile. They have a few cristae projecting into a dense matrix. Opaque matrix granules are not common in neuronal mitochondria. Microtubules (neurotubules) and neurofilaments are abundant in the cytoplasm. These structures tend to run in tracts around the nucleus and around concentrations of other organelles, such as Nissl bodies, and then eventually funnel into the nerve processes.

The axon (Ax) arises from a cone-shaped region of the perikaryon known as the axon hillock. In large neurons, the axon hillock contains a diminished number of rough-

(Continued)

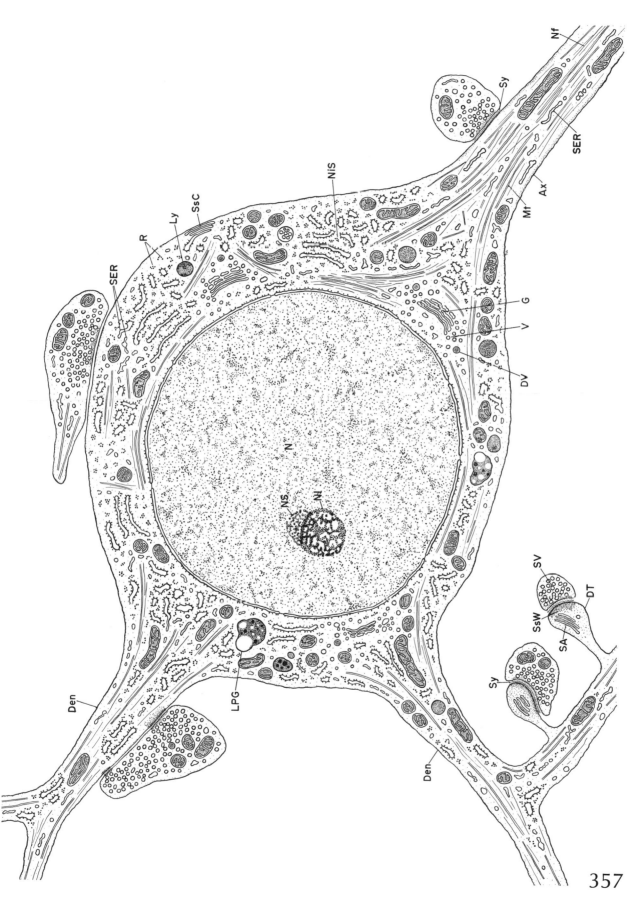

157—NEURON (*Continued*)

surfaced cisternae and ribosomes. Bundles of microtubules (Mt) course from the perikaryon into the axon. In the initial segment of the axon there is a layer of dense material applied to the inner surface of the plasma membrane. This dense undercoating, which is similar to that occurring in the node of Ranvier (Plate 159), extends to the beginning of the myelin sheath in myelinated axons. Neurofilaments (Nf) and elongated tubules of smooth endoplasmic reticulum (SER) also run into the axon. Mitochondria are elongated and their cristae may extend longitudinally. A few vesicles may be seen in the axoplasm.

Dendrites (Den) can be considered extensions of the perikaryon and, unlike axons, contain the same organelles as the perikaryon. Several dendrites may emerge from the cell body. They may branch and are always unmyelinated. Elements of rough-surfaced endoplasmic reticulum and ribosomes occur in the dendrites but become less prominent distally. Microtubules are common, but neurofilaments are not as abundant. Mitochondria, channels of smooth endoplasmic reticulum, vesicles, and lysosomes occur as well.

Spines or thorns (dendritic thorns, DT) project from the surfaces of the dendrites. These are most common after the bifurcation of the main dendrite and consist of a narrow neck and a dilated terminal bulb. The cytoplasm of the spine often contains a parallel array of flattened cisternae known as the spine apparatus (SA). A band of dense material occurs in the cytoplasm between cisternae. Microtubules and neurofilaments do not extend into the spines, which, instead, contain finely fibrillar or flocculent material.

Axon terminals from synapses (Sy) with all parts of the neuron: the dendrites, either spines or dendritic trunks (axo-dendritic); the cell body (axo-somatic); and the axon (axo-axonal). In general, the synapse consists of an enlargement of the axon, which contains mitochondria and many synaptic vesicles (SV). The vesicles tend to be clustered toward the presynaptic membrane. Most of the vesicles are small (400 to 500 Å) with clear contents. Larger (600 to 800 Å) dense core vesicles are common in the endings of autonomic nerves. The pre- and postsynaptic membranes are separated by a cleft of 200 to 400 Å, which contains moderately dense material. A dense line may occur in the material somewhat closer to the postsynaptic membrane. Dense material is applied to the inner side of both junctional membranes. Fine filamentous material forming a subsynaptic web (SsW) may be associated with the postsynaptic membrane.

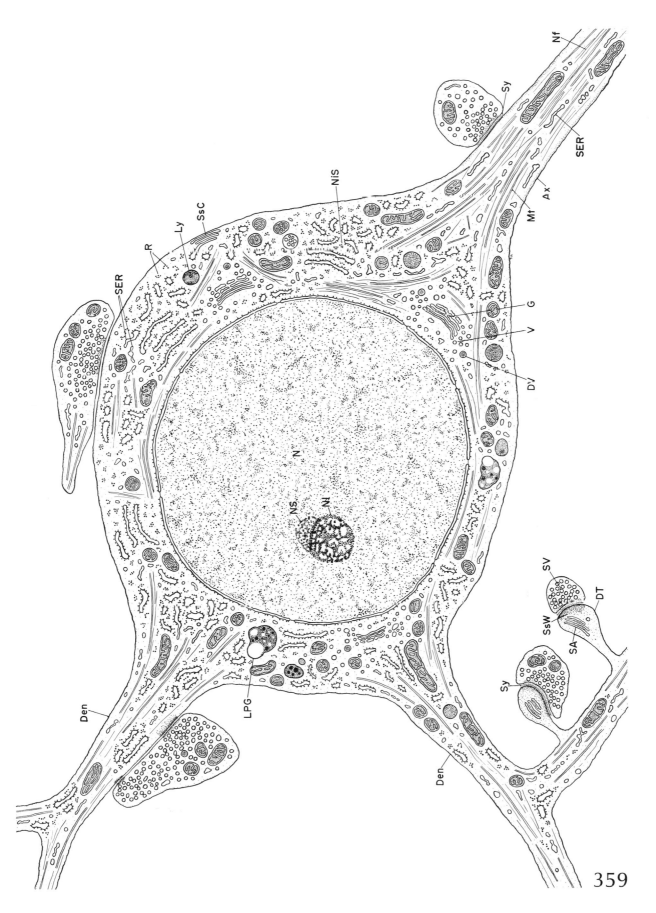

158—MYELINATED NERVE AND SCHWANN CELL

The myelinated nerve consists of an axon enclosed by the Schwann cell (SC) and by concentric lamellar specializations of the Schwann cell plasma membrane (PM) that form the myelin sheath (MS). The axon is bounded externally by a plasma membrane (PM) and is composed of axoplasm of low density. Neurofilaments (Nf), microtubules (Mt) (neurotubules), and channels of smooth endoplasmic reticulum (SER) extend longitudinally down the axon. Neurofilaments are numerous and have a diameter of about 85 Å. Microtubules, not as abundant, have a larger diameter (220 Å) and a lucent, central core. Tubular channels of smooth-surfaced endoplasmic reticulum pursue a sinuous course down the axon. These channels are of irregular diameter and may have focal expansions along their lengths. Also occurring in the axoplasm in small numbers are mitochondria, small vesicles (V), and dense core vesicles (DV).

The Schwann cell envelops the nerve fiber. The nucleus is usually located peripherally and is oval or crescent-shaped. The cytoplasm contains a Golgi apparatus, mitochondria, a few cisternae of endoplasmic reticulum, and some ribosomes. Pinocytotic vesicles are associated with the plasmalemma of the Schwann cell. The cell is enveloped by a basement membrane or external lamina (BL).

The myelin sheath is a spiral wrapping of Schwann cell plasma membrane around the axon. The plasma membrane (PM) is a three-layered structure (unit membrane) consisting of two dense lines separated by a clear zone. Each layer is approximately 25 Å thick. At a point on the surface of the Schwann cell, the plasma membrane is infolded. The opposing membranes and intervening channel are known as the mesaxon (Ma). The outer surfaces of the unit membrane fuse to form the thin intraperiod or intermediate lines of the myelin sheath. The inner or cytoplasmic surfaces of the plasma membrane fuse to form a thicker line, the major dense line. The fused membranes are wrapped spirally for several turns around the axon. Internally, the membranes separate, form an internal mesaxon, and diverge to surround the axon.

An early stage in myelin formation is illustrated at the lower left of the Plate. The mesaxon spirals around the axon embedded in the Schwann cell, but the mesaxonal turns are few in number and still separated by Schwann cell cytoplasm. In nerve conduction (see Plates 159, 160) the myelin sheath acts as an insulator, preventing the spread of an impulse to adjacent axons. Conduction is more rapid in myelinated than in unmyelinated fibers. It is also faster in fibers of larger diameter and with thicker myelin sheaths.

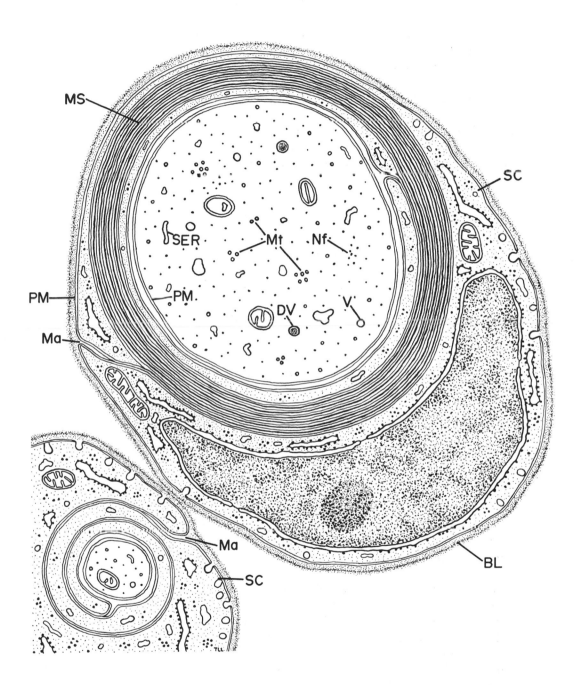

159—MYELINATED NERVE—NODE OF RANVIER

The nodes of Ranvier are periodic interruptions in the myelin sheath (MS) of the peripheral nerve processes. Each segment of the sheath demarcated by the interruptions is composed of a single Schwann cell (SC). Where the two adjacent Schwann cells abut one another at the node, the myelin sheath is absent, but villi and projections of the Schwann cells interdigitate. Near the node, on the axonal side of the sheath, each lamella separates from the compact myelin to form a mesaxon (Ma). The mesaxon extends to the surface of the axon and splits into its component plasma membranes (PM), each of which is continuous with a membrane derived from the adjacent mesaxon. Schwann cell cytoplasm occupies the folds formed by the splitting of the myelin sheath. Dense granules, 100 to 150 Å in diameter, are found in the Schwann cell cytoplasm nearest the axon surface.

The axon narrows as the node is approached, and then widens slightly at the node. The axon membrane comes into close proximity to the Schwann cell folds. Dense material may accumulate beneath the axolemma in these regions and in the nodal region. In the paranodal regions, short thickenings occur periodically on the outer surface of the axolemma. These thickenings form rings that encircle the axon. In the axoplasm are found mitochondria, neurofilaments (Nf), neurotubules (microtubules, Mt), and channels of smooth-surfaced endoplasmic reticulum (SER).

The inside of the nerve fiber has a resting potential of -90 mv with respect to the outside. This potential is maintained by the active pumping out of sodium from the axon. When the membrane is stimulated, depolarization occurs and the potential inside becomes positive. Depolarization is brought about by a change in the permeability of the membrane which allows sodium to enter. Repolarization occurs when the membrane becomes impermeable to sodium but permeable to potassium, which then leaks out of the cell. The voltage change occurring during the polarization-repolarization process is called the action potential.

Conduction in myelinated nerves is extremely rapid and has been explained by saltatory conduction of the action potential between the nodes of Ranvier. According to this theory, the threshold for stimulation is much less at the node than in the internodal regions. Since an electric field is more likely to evoke a change in potential or depolarization in a region of low threshold (node) than in one of high threshold (internode), the action potential skips from node to node. Structurally, the axon membrane at the node differs from the membrane elsewhere in that it has a dense layer beneath it. It is not known whether or not this layer affects the excitability of the membrane, but similar material is found in the initial axon segment (see Plate 157), which has a lower threshold for excitation and which is the site where the action potential originates. The thickenings on the outer surface of the axolemma, furthermore, delimit the nodal region from the internodal areas. These thickenings could obstruct the flow of ions between the extracellular space surrounding the node and the space between the internodal myelin sheath thereby preventing spread of excitation from the nodal to the internodal regions.

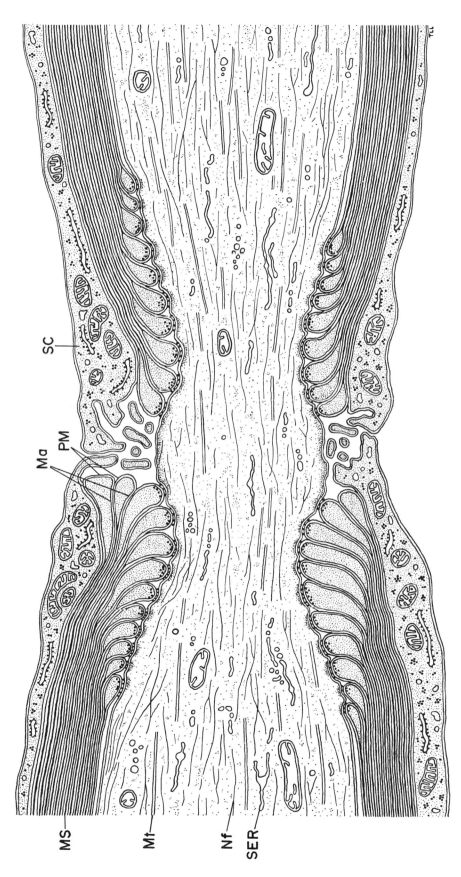

160—UNMYELINATED NERVE

Unmyelinated fibers are enclosed by Schwann cells (SC), as are the myelinated nerves. Several axons (Ax) may be embedded in a single cell. The plasma membrane of the Schwann cell folds in from the surface near each fiber to form a mesaxon (Ma), and envelops the nerve fiber. However, in unmyelinated nerves the Schwann cell membrane does not form a spiral multilayered wrapping around the nerve fiber.

Both the structure of the Schwann cell cytoplasm and the axoplasm of the nerve fibers are similar to those of myelinated nerves. The unmyelinated fibers are of smaller diameter but contain all of the structures (neurofilaments, microtubules, channels of endoplasmic reticulum, mitochondria, vesicles, and dense core vesicles) found in myelinated fibers. The surface of the Schwann cell is covered by a basement membrane. The nucleus is found in the center of the cell.

In unmyelinated nerves, the membrane becomes depolarized at the point of stimulation in the same manner as myelinated nerves. The propagation of the nerve impulse is explained by the local circuit theory. The surface of the active region is negative with respect to the inactive region, so that current flows from the inactive to the active region. Inside the membrane, the active region is more positive with respect to the adjacent inactive region, so that current flows from active to inactive regions. A local circuit is established, which excites the adjoining inactive region, which excites the next point, and so on. Thus, local current flow and successive depolarizations ahead of the action potential are responsible for conduction of the impulse along the nerve fiber. In unmyelinated nerves, conduction is a continuous process and is slower than the saltatory conduction of myelinated nerves.

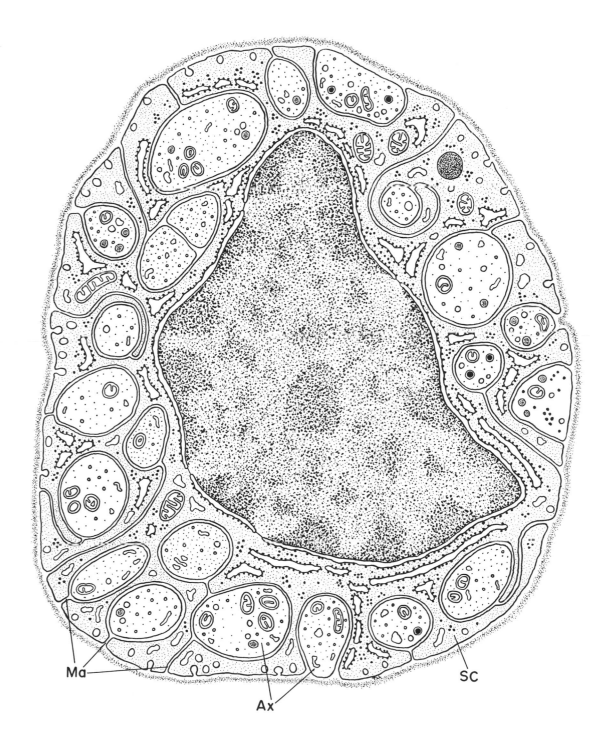

Ma

Ax

SC

NERVOUS SYSTEM

161 — PROTOPLASMIC ASTROCYTE

Neuroglial cells occupy most of the interstitial space between neurons of the central nervous system. With the light microscope, they have been divided into two categories: the macroglia and the microglia. The macroglia have an ectodermal origin and consist of astrocytes and oligodendrocytes. The microglia are thought to be of mesodermal origin and are small cells that become phagocytic in response to injury or disease. With electron-microscopic observation, however, no cells corresponding to the microglial cells of light-microscopic preparations have been identified with certainty until recently (Plate 164). It has been suggested also that microglia have been included within the category of oligodendrocytes or that they do not actually exist as a separate cell type, but instead are macrophages derived from perivascular adventitial cells or from other components of the reticuloendothelial system. Confusion also existed in the identification, with electron-microscopic observation, of the other neuroglia, but these have now been sufficiently characterized to be identified with some confidence.

Astrocytes are star-shaped cells with cytoplasmic processes extending into the neuropil or terminating in end feet (EF) on the surfaces of blood vessels. Protoplasmic astrocytes are found in the grey matter and have fewer cytoplasmic fibrils than do the fibrous astrocytes of the white matter. The protoplasmic astrocyte has an irregular outline with a round or oval nucleus. The chromatin material of the nucleus is dispersed. The perinuclear cytoplasm contains a small Golgi apparatus, scattered short cisternae of rough-surfaced endoplasmic reticulum, and free ribosomes.

Mitochondria, glycogen granules (Gly), and a few microtubules are scattered in the cytoplasm. The hyaloplasm is of relatively low density. The cytoplasmic filaments (Fl) occur in tightly knit bundles or fibrils that extend parallel to the long axis of the cell and into the processes.

Processes of varying size extend between the axons and dendrites of the neuropil. They may be attenuated sheets or of large diameter. The small processes contain ribosomes, glycogen granules, and fibrils, while the larger processes may contain other organelles. Some of the processes terminate in end feet (EF) that surround the capillaries of the central nervous system (Plate 43). The terminals are bounded by a basal lamina (BL) and have localized densities applied to the inner aspect of the plasma membrane.

Bundles of nerve fibers are enveloped by the astrocyte processes (upper left of Plate). The axons (Ax) are small, and transverse section shows them to contain a few microtubules. The dendrites (Den) are larger and, unlike axons, may contain also elements of endoplasmic reticulum and ribosomes. Axons synapse (Sy) on the dendrites or dendritic thorns (DT); the latter are identified by the presence of a spine apparatus (SA). The axon terminal contains large numbers of synaptic vesicles (SV) (see also Plate 157).

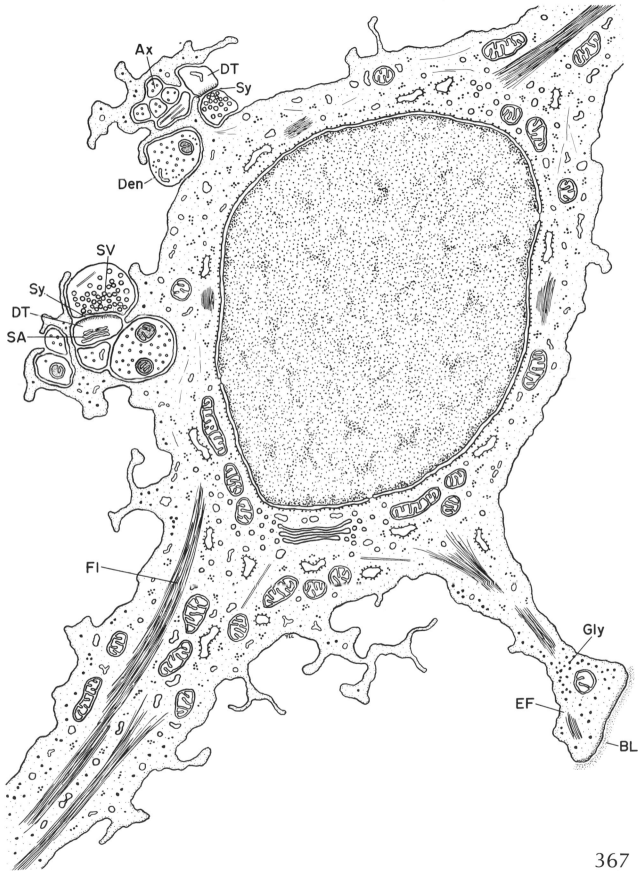

162—FIBROUS ASTROCYTE

Fibrous astrocytes occupy the white matter and can be distinguished from the protoplasmic astrocytes. A number of broad processes arise from the cell bodies of fibrous astrocytes. The large processes then branch to form thinner processes. In contrast to the protoplasmic astrocytes, the surface of the fibrous astrocytes is smoother in contour and not as deeply indented by adjacent nerve processes. The most prominent component of the cytoplasm is the extensive system of filaments (Fl). They extend into the processes but, unlike protoplasmic astrocytes, are not organized into tight bundles and, instead, are more dispersed.

The central nucleus has an irregular or indented outline and contains dispersed chromatin. The hyaloplasm is of low density with a sparse population of organelles. The Golgi apparatus is small and elements of the endoplasmic reticulum exist as short, scattered cisternae. Mitochondria are often elongated. Other structures present are lysosomes, free ribosomes, and glycogen granules. A pair of centrioles or even a cilium is sometimes found. End feet of fibrous astrocytes, as well as of the protoplasmic astrocytes, terminate on the surface of capillaries. In addition, the end feet also form a layer, the glia limitans, at the surface of the neural tissue between the neurons and pial elements.

Several functions have been attributed to the glial cells. First, the positional relationship of glial cells to neurons indicates they may provide structural support to the nerve tissue. The glial cells also respond to injury to the central nervous system by proliferation and scar formation. The relationship of the astrocyte end feet to capillaries suggests a role in regulation of transfer of substances from the capillaries to the nerve tissue. The actual barrier to some substances, though, seems to reside in the capillary endothelial cells (Plate 43). It has been suggested that neuroglial cells have an isolating role in the central nervous system in that they separate the receptive surfaces of neurons (perikaryon and dendrites) from nonspecific afferent influences (Palay and Peters). In this way, the axon terminals act in a discrete and localized manner.

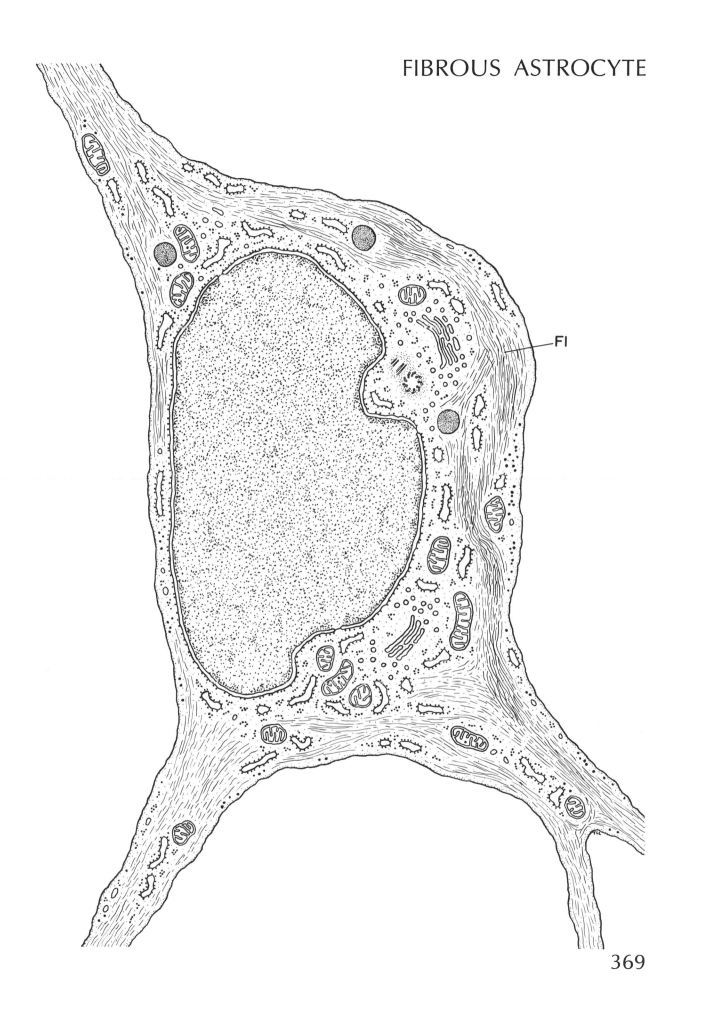

Fl

163 — OLIGODENDROCTYE

Oligodendrocytes are found in the white matter between nerve fibers (interfascicular oligodendrocytes), and in the gray matter where they are also associated with nerve fibers. In the gray matter, some oligodendrocytes are additionally situated in close association with neuronal perikarya and are called satellite cells. Oligodendrocytes are small and have only a few processes extending from the cell body. The most conspicuous differences between these cells and astrocytes are the dense hyaloplasm and the clumping of nuclear chromatin. The nucleus is oval and located at one pole of the cell; it contains large masses of condensed chromatin, especially adjacent to the nuclear envelope. The Golgi apparatus (G) occupies a large portion of the cell. Cisternae of rough-surfaced endoplasmic reticulum are common and are usually flattened. Free ribosomes are also abundant in the cytoplasm, but only a small number of round or oval mitochondria and dense bodies or lysosomes occur. In contrast to astrocytes, oligodendrocytes possess few cytoplasmic filaments or glycogen granules. Microtubules (Mt), on the other hand, are present in large numbers and course into the processes.

Two major functions have been attributed to oligodendrocytes. First, it is generally thought that the myelin sheath (MS) of the central nervous system is formed by oligodendrocytes. In adult animals, however, few examples have been reported in which the plasma membrane of oligodendrocytes and the myelin sheath have been seen to be clearly continuous. During development, the glial processes wrap around the axon (Ax), and the membranes become apposed and spiral around the axon, forming the lamellae of the myelin sheath (unit membrane structure illustrated in region of myelination; see also Plate 158). The only difference between peripheral and central nervous system myelin is that in the latter most of the cytoplasm is squeezed out of the outside of the sheath. In the peripheral myelin sheath, the cytoplasm forms a complete layer around the sheath, but in central myelin the cytoplasm is restricted to the narrow tongue process (TP). In both, the intermediate dense line is formed by the apposition of the outer surfaces of the membrane when it turns inward, and the major dense line is formed by the approximation of the cytoplasmic faces of the unit membrane. The second function attributed to oligodendrocytes is a role in the maintenance and nutrition of the neurons they enclose.

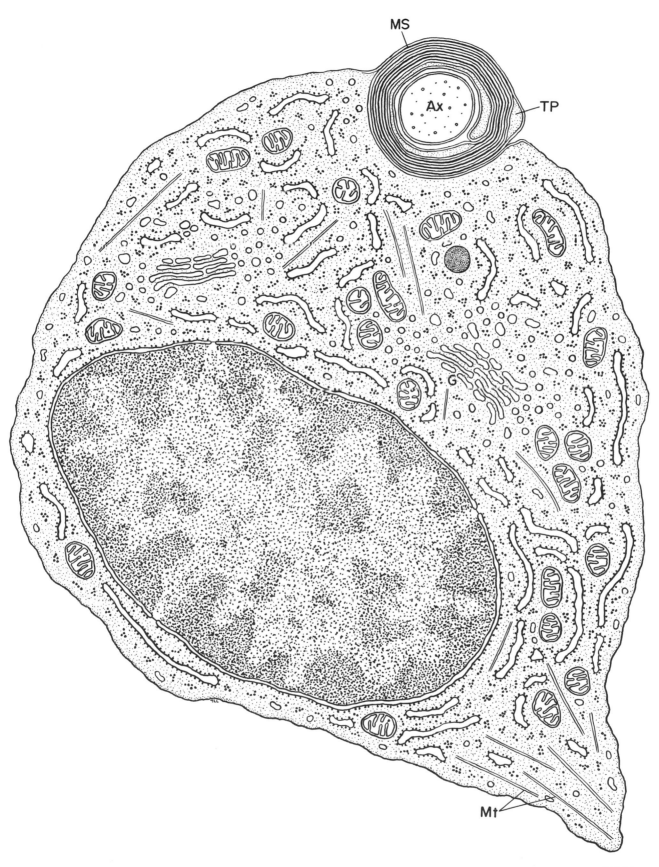

164—MICROGLIA

Of the glial cells, microglia have presented the greatest difficulty in identification at the fine structural level. However, with the use of a specific metallic stain for microglia (weak silver carbonate method of del Rio-Hortega), these cells have been identified with the electron microscope (Mori and Leblond). Microglia have been observed in the brain tissue proper (interstitial microglia) and associated with capillaries (pericytal microglia). The small neuroglial cell or third neuroglial type (Vaughn and Peters) is similar in structure to the microglia but may represent a multipotential glial element of neuro-ectodermal origin.

Microglia, which are smaller than other glial cells, are usually ovoid in profile with a number of cytoplasmic protrusions and processes. The nucleus is also oval and contains a nucleolus and prominent masses of clumped chromatin, especially adjacent to the nuclear envelope. The density of the hyaloplasm is intermediate between that of astro-cytes and oligodendrocytes. A Golgi zone, sometimes extensive, occurs near the nucleus. A small number of cisternae of rough-surfaced endoplasmic reticulum are present. These may be elongated, slightly dilated, and filled with moderately dense material. Free ribosomes are also common. Mitochondria are round or oval and have a dense matrix containing opaque granules. In contrast to other glia, microglia are largely devoid of filaments, glycogen granules, or microtubules.

The most characteristic feature of microglia is their population of dense bodies. The smaller dense bodies appear to originate in the Golgi complex and correspond to lysosomes. Larger bodies resemble lipofuscin pigment granules. A few lipid droplets may occur as well.

Microglia have similarities with the macrophages and pericytes of the reticuloendothelial system. Their abundance of lysosomes might be indicative of a phagocytic function. These observations support the theory that under conditions of inflammation or injury, microglia become motile and phagocytic, functioning as macrophages in the brain.

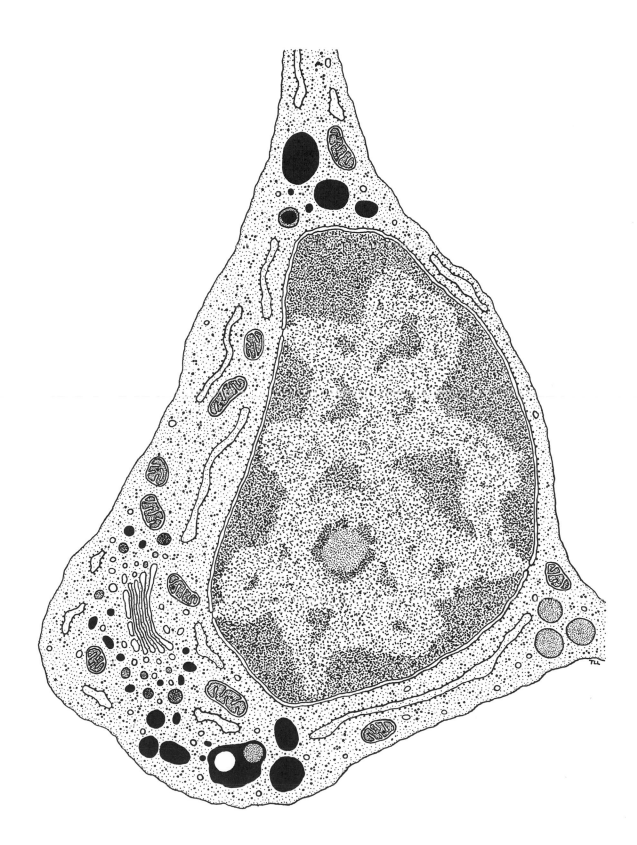

165 — EPENDYMAL EPITHELIAL CELL

The ependyma lines the ventricles of the brain and the central canal of the spinal cord. The cells are cuboidal to columnar in shape and have a large, round, basal nucleus with a nucleolus. The apical surface is provided with many cilia (C) and villous processes. The cilia are reduced in diameter toward their distal ends and in these regions contain only nine single ciliary fibers. Proximally, they terminate in basal bodies (BB) composed of nine triplet fibers. Filamentous appendages or rootlets (Rt) extend laterally and basally from the basal bodies, and in some species these rootlets are striated. Irregular villi and cytoplasmic processes project from the surface between the cilia. The lateral cell borders are relatively straight and adjacent cells are joined by junctional complexes. The basal surface of the ependymal cells rests directly on the underlying nervous tissues.

A small Golgi apparatus is located above the nucleus. The endoplasmic reticulum consists of short, rough-surfaced cisternae scattered throughout the cytoplasm. Free ribosomes are abundant and often occur in clusters and rosettes. Smooth cisternae and vesicles are distributed throughout the cytoplasm. Mitochondria are elongated and most abundant in the apical cytoplasm. Multivesicular bodies and dense bodies or lysosomes are also common, and some contain small dense particles resembling ferritin. Running between the organelles is an extensive system of cytoplasmic filaments (Fl). These are most extensive around the nucleus, where they may form large bundles. Substances can pass from the ventricular fluid into the brain parenchyma, and ependymal cells probably play a role in the regulation of their transfer.

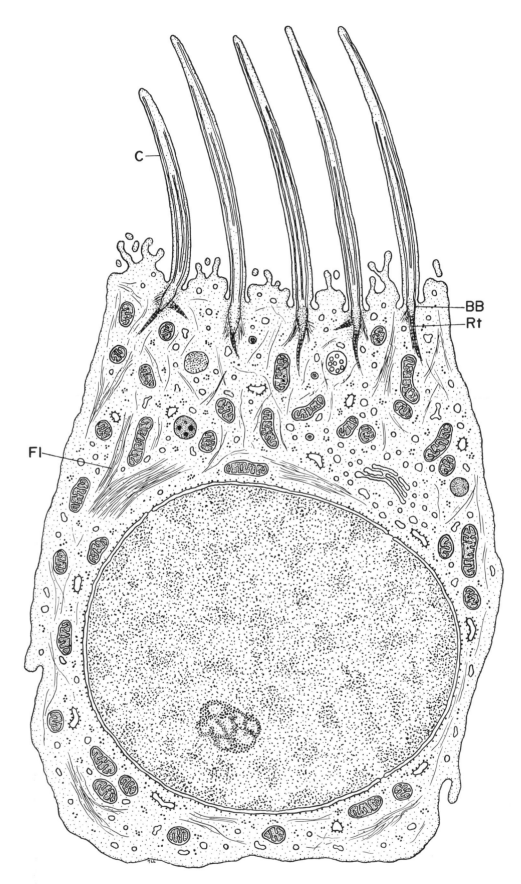

166—CHOROID PLEXUS EPITHELIAL CELL

The choroid plexuses are found in the walls of the lateral, third, and fourth ventricles of the brain. They consist of a layer of epithelium overlying highly vascular pia mater. The epithelium is a modification of the ependymal lining of the ventricles. The epithelial cells of the choroid plexus are cuboidal with a basally situated nucleus containing a nucleolus. The rounded apical surface is provided with a number of bulbous or clavate microvilli. The lateral cell borders are relatively straight, but basally a few infoldings may occur. A small Golgi apparatus occurs above the nucleus. Cisternae of rough-surfaced endoplasmic reticulum are short but fairly common throughout the cytoplasm. Clusters of ribosomes occur free in the cytoplasm. Rod-shaped mitochondria with dense matrices are abundant and a few lysosomes are seen. Filaments run in all directions in the cytoplasm between organelles. The choroid plexuses are the chief source of cerebrospinal fluid, transferring fluid from the capillaries to the ventricles. The amplification of basal and apical plasma membranes is similar to that seen in other cells transporting fluids. The blood-brain barrier does not exist in the choroid plexus and the capillaries in this region are fenestrated, unlike those elsewhere in the brain (see Plate 43).

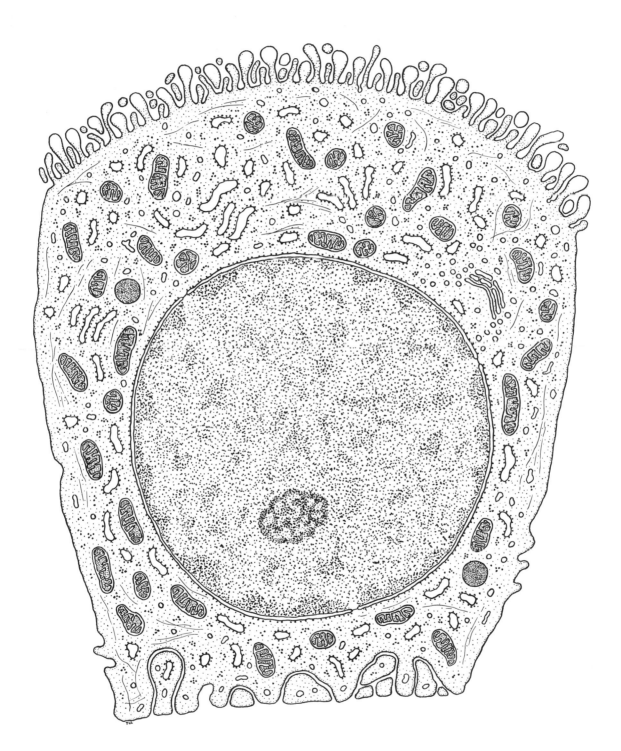

REFERENCES

Blinzinger, K. and H. Hagar, 1962. Elektronenmikroskopische Untersuchungen über die Feinstruktur ruhender und progressiver Mikrogliazellen im Säugetiergehirn. Beit. Path. Anat., 127:173–192.

Brightman, M. W. and S. L. Palay, 1963. The fine structure of ependyma in the brain of the cat. J. Cell Biol., 19:415–439.

Elfvin, L.-G., 1961. The ultrastructure of the nodes of Ranvier in cat. sympathetic nerve fibers. J. Ultrastruct. Res., 5:374–387.

Elfvin, L.-G., 1961. Electron microscopic investigation of the plasma membrane and myelin sheath of autonomic nerve fibers in the cat. J. Ultrastruct. Res., 5:388–407.

Hirano, A., 1968. A confirmation of the oligodendroglial origin of myelin in the adult rat. J. Cell Biol., 38:637–640.

Klinkerfuss, G. H., 1964. An electron microscopic study of the ependyma and subependymal glia of the lateral ventricle of the cat. Amer. J. Anat., 115:71–100.

Kruger, L. and D. S. Maxwell, 1966. Electron microscopy of oligodendrocytes in normal rat cerebrum. Amer. J. Anat., 118:411–436.

Mori, S. and C. P. Leblond, 1969. Identification of microglia in light and electron microscopy. J. Comp. Neur., 135:57–80.

Peters, A., S. L. Palay, and H. de F. Webster, 1970. The Fine Structure of the Nervous System. The Cells and Their Processes. Hoeber, New York, 198 pp.

Revel, J.-P. and D. W. Hamilton, 1969. The double nature of the intermediate dense line in peripheral nerve myelin. Anat. Rec., 163:7–16.

Robertson, J. D., 1960. The molecular structure and contact relationships of cell membranes. Prog. Biophys., 10:343–418.

Vaughn, J. E. and A. Peters, 1968. A third neuroglial cell type. An electron microscopic study. J. Comp. Neurol., 133:269–288.

EYE

167—ROD CELL

There are two main types of photoreceptors in the retina: the rod cells and the cone cells. These are long, slender cells with the outer or apical portion facing the pigment epithelium and the opposite end facing the inner or vitreous pole of the retina. The cells have been divided into several compartments. The outer segment (OS) is characterized by its content of parallel membranous lamellae. The lamellae are oriented transversely to the long axis of the cell and are formed by closely packed, flattened, membranous sacs. In some central regions, the sacs are perforated, producing vesicular profiles. In rod cells, the lamellae are enclosed by the limiting membrane of the outer segment. The outer segment is connected to the inner segment (IS) by a slender stalk. The stalk contains a modified cilium (C) that lacks the two central fibrils and that originates from a basal body (BB) in the distal end of the inner segment. A centriole (Ce) is situated nearby. A striated rootlet (Rt) extends downward from the basal body.

The inner segment is subdivided into an outer ellipsoid and an inner part or myoid. The ellipsoid contains a large number of mitochondria (M). These are elongated and contain many transverse cristae. The myoid, on the other hand, contains a large Golgi apparatus with many associated small vesicles. Some larger vacuoles containing flocculent material occur nearby. Short ribosome-studded cisternae as well as large numbers of free ribosomal clusters occur in this region. Channels of smooth-surfaced endoplasmic reticulum and microtubules are also found.

The outer fiber (OF) extends from the inner segment to the cell body. At the junction of the inner segment with the rod fiber, a junctional complex is formed with adjacent Müller cells (Plate 169). This zone corresponds to the external limiting membrane of the retina. The thin rod fiber contains primarily smooth endoplasmic reticulum and ribosomes. The cell body contains the small, oval nucleus.

The basal cytoplasm is attenuated into the rod axons or inner fibers (IF), which terminate in the bulb-shaped rod spherules (RS). Membranous channels and microtubules course down the axon. The spherule contains large numbers of small vesicles thought to be synaptic vesicles (SV). Mitochondria and elements of endoplasmic reticulum are found in smaller numbers. The spherule is indented by the terminations of bipolar neurons and by processes of horizontal neurons. A gap of about 200 Å exists between the nerve terminals and rod spherule. There are no prominent thickenings of the plasma membranes. Vesicles are found in the nerve endings of horizontal cell processes that are closest to the spherule. The less deeply inserted processes of the bipolar cells contain fewer vesicles. Synaptic ribbons or bars (SB) (see also Plate 174) occur in the rod cytoplasm of the spherule. These structures are dense rods situated perpendicular to the plasma membrane where it protrudes slightly opposite the cleft between two nerve fibers. The dense rod is enclosed by vesicles.

The outer segments of the rods contain the visual pigment rhodopsin, which is thought to reside in the membranous lamellae. Rhodopsin consists of retinene, an aldehyde of vitamin A, conjugated with the protein opsin. Upon exposure to light, the rhodopsin is isomerized to lumirhodopsin, which is unstable and yields metarhodopsin. Metarhodopsin is hydrolyzed to retinene and opsin. It is thought that visual excitation is triggered in the initial step of the transformation of rhodopsin to metarhodopsin. Depolarization of the receptor cell membrane is conducted over the cell body and axon, leading to the initiation of impulses in the nerve terminals at the base of the cell.

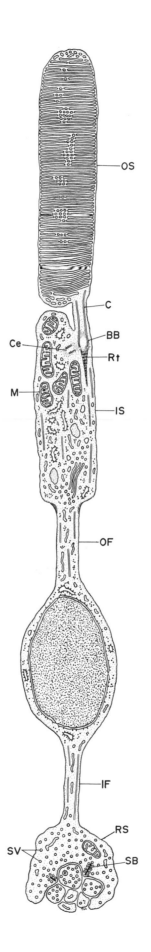

OS

C

BB
Ce
Rt

M

IS

OF

IF

RS
SV
SB

168 — CONE CELL

Cone cells are composed of the same basic parts as the rod cells, but there are some important differences in certain details. The outer segment (OS) has a broader base and tapers toward the tip. As in rods, it is composed of flattened membranous sacs stacked like coins. Unlike rods, however, many of the sacs are continuous at their ends with the limiting membrane of the outer segment, especially toward the base of the outer segment. The inner segment (IS) is larger and the ellipsoid portion is packed with elongated, longitudinally oriented mitochondria. The myoid contains large numbers of ribosomes, which cause the cytoplasmic basophilia observed with the light microscope. The outer fiber (OF), when present, is short in cones. The nucleus of the cell body is larger than that of rods and the chromatin is not as highly condensed.

A thick inner fiber (IF) or axon terminates in a broad pedicel containing synaptic vesicles (SV). Groups of three nerve terminals are found indented into the base of the pedicel. The contacts are similar in structure to those on the rod spherule. The two lateral terminals of the triad contain vesicles and are thought to represent processes of the horizontal cells. The central fiber, which does not contact the pedicel, contains profiles of endoplasmic reticulum, ribosomes, and microtubules, and it probably represents the dendrite of bipolar cells. The synaptic ribbon or bar (SB) is situated in the pedicel opposite the interface of the two horizontal cell endings. Other bipolar cells terminate on the surface of the pedicel, but these contacts exhibit little synaptic specialization. Processes of the pedicel sometimes contact rod spherules (RS). In the region of contact, which most likely has an adhesive function, there is a symmetrical accumulation of dense material on the inner sides of the plasma membranes. Shorter zones of close membrane apposition have also been described.

Cones are responsive to light of different wave lengths and thus are capable of distinguishing colors. There appear to be three groups of cones: red-, green-, and blue-sensitive. Similarly, there may be three types of color vision pigments, although each type of cone cell could contain more than one pigment. Cones are best suited for seeing in bright light, while rods have a greater sensitivity for weak light stimuli and are better adapted for vision in dim light.

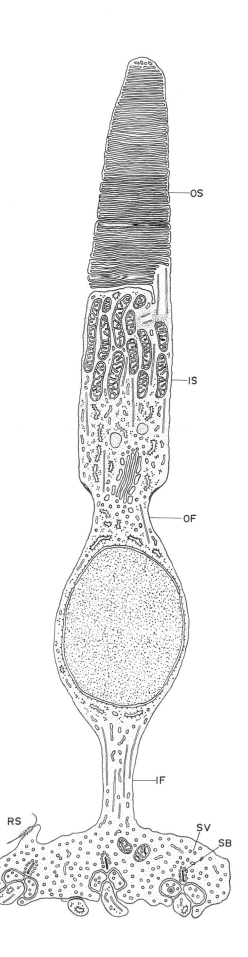

OS

IS

OF

RS

IF

SV

SB

169—MÜLLER CELL

Müller cells are the supporting elements of the retina and have many cytoplasmic features comparable to those of glial cells. They are highly columnar cells extending from the inner surface of the retina to beyond the external limiting membrane between the inner segments of the receptor cells. The cytoplasm of the Müller cells encloses, or is interposed between, nerve cells, dendrites, and axons. The nuclei (N) are found at the level of the inner nuclear layer of the retina. The cells can be divided into a basal portion, extending from the nucleus to the internal limiting membrane, and an apical or outer half on the other side of the nucleus.

The base is broadened and the surface is somewhat irregular. A thick basement lamina (BL) underlies the cells and constitutes the internal limiting membrane of the retina. The basal cytoplasm contains large numbers of cytoplasmic filaments (Fl). Elongated mitochondria are oriented in the long axis of the cell. Vesicles and smooth tubules are common in the cytoplasm. There are also a few small dense granules. Rough-surfaced endoplasmic reticulum is not prominent and consists of some short cisternae near the nucleus. Free ribosomes are also found in this area. The same structures are found in the apical portion of the cell, but they are not as densely packed, thereby giving the cytoplasm a lighter appearance. A Golgi apparatus occurs near the nucleus. The cells extend between the rods and the cones and, with them, form junctional complexes that make up the external limiting membrane. Beyond the external limiting membrane, the cells terminate in thin villous processes.

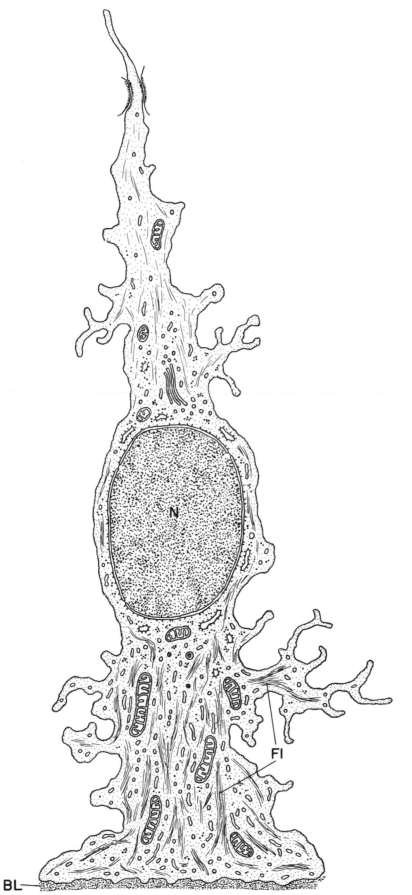

170—PIGMENT EPITHELIAL CELL

The pigment epithelium has traditionally been considered as a layer of the retina. It may more logically, however, belong to the choroid, because the basement lamina (BL) of the pigment epithelium is part of the glassy or Bruch's membrane of the choroid. The cells of the pigment epithelium are cuboidal and have a basal nucleus. The apical surface has a large number of long villous processes. These are larger than microvilli, and they project into the interstices between the outer segments of the rods and cones. The basal surface of the cell bears a number of short infoldings and rests on a basement lamina. Besides the elaborate surface processes, the cells are characterized by pigment granules (PiG). These are round or oval and considerably larger than those observed in melanocytes (Plate 53). They appear to originate, however, in the same manner from lamellated melanosomes. Most of the pigment granules occur in the apical half of the cell. The cells also possess a fairly extensive smooth-surfaced endoplasmic reticulum (SER); but rough-surfaced cisternae are not nearly as abundant, nor are free ribosomes common. Large mitochondria with transverse cristae congregate in the bottom half of the cell. A Golgi complex and a few small dense granules also occur in the basal regions.

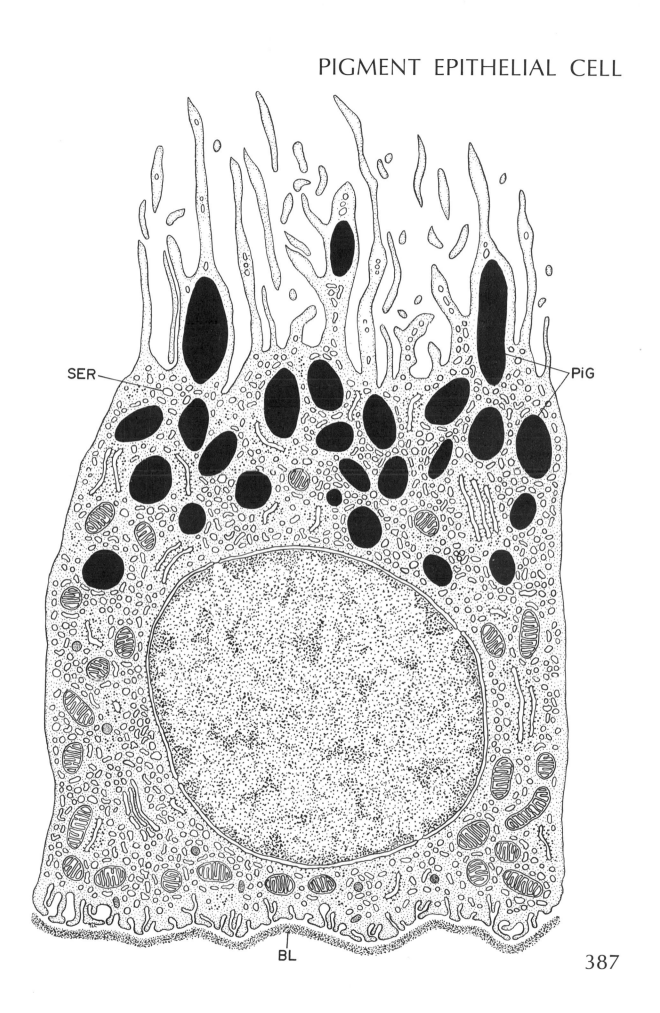

SER

PiG

BL

171 — CILIARY EPITHELIAL CELL

The ciliary epithelium consists of two cell layers: an outer layer of pigmented cells and an inner layer of non-pigmented cells. The outer layer is continuous with the pigment epithelium of the choroid (Plate 170). It continues onto the posterior surface of the iris where the cells lose their pigment and are transformed into the myoepithelium of the dilator pupillae. The non-pigmented ciliary cells are continuous posteriorly with the sensory portion of the retina and anteriorly with the pigment epithelium on the posterior surface of the iris.

When compared with other epithelial cells, the non-pigmented ciliary epithelial cells appear to be upside down. The free surface of the cell faces the posterior chamber and would normally be considered apical. However, this surface is lined by a basement lamina (BL) and its structure is similar to that of the basal surfaces of other epithelia. The opposite pole of the cell more closely resembles an apical region but faces the layer of pigmented cells. This reversal of structural polarity results from the invagination of the optic vesicle to form the optic cup during development. As a result, the basal surface becomes superficial, while the apical surface is applied against the pigmented layer.

The most striking feature of the outer epithelial cells is the presence of deep infoldings of the free (basal) surface. These appear as deep folds of the plasma membrane and occur on the lateral surfaces as well. These infoldings have been shown to be produced by interdigitation of slender cytoplasmic processes of adjacent cells. Small *maculae adherentes* (MA) are found on some of the interdigitations.

The opposite surface of the cell is not nearly so irregular in contour and bears only a few short microvilli. Junctional complexes are found along the lateral surfaces near the apex. A Golgi apparatus occurs in the cytoplasm and has a relatively large number of small vesicles associated with it. Some dense bodies resembling lysosomes are seen nearby. There are a few short cisternae of rough-surfaced endoplasmic reticulum and free ribosomes. A number of smooth-surfaced membranes and vesicles are seen. Arrays of vesicular profiles occur at the free surface in osmium-fixed material, but these are thought to be artifacts produced by breakdown of the interdigitations. Mitochondria are common in these cells and are round to oval with transverse cristae. The nucleus is central in position, contains a nucleolus, and is infolded.

The ciliary epithelium is responsible for the production of the aqueous humor. This is thought to occur as a result of active transport of water and ions across the epithelium. Both the folding of the plasma membrane and the presence of large numbers of mitochondria are characteristic of cells engaged in active transport of salts and fluids (see also Plates 55, 68, 101, 176, 182).

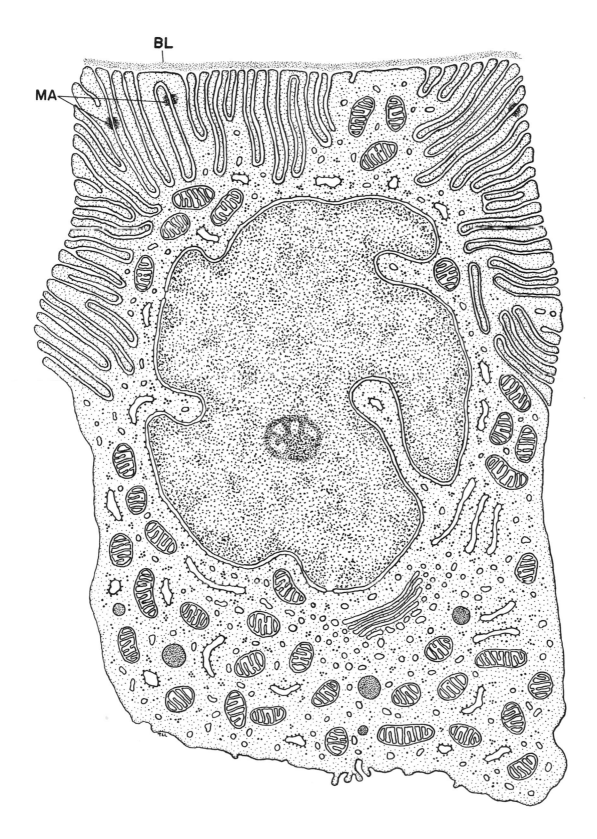

172—LENS EPITHELIAL CELL

A layer of cuboidal epithelial cells underlies the capsule of the lens on the anterior surface. Toward the equator, the cells become elongated in the process of differentiation into lens fibers. The cells are displaced inward, forming the lens bow radiating from the equator to the center or nucleus of the lens. The cells in the bow area are elongated and represent newly formed lens fibers. Knob-like protrusions occur along the cell surface and indent adjacent cells. Tight junctions are found along these areas, and desmosomes occur elsewhere. The cells have a central nucleus with a nucleolus and dispersed chromatin. The bulk of the hyaloplasm is filled with moderately dense, fine, granular material. Organelles are very few in number. Some small Golgi elements, short cisternae of endoplasmic reticulum, scattered clusters of ribosomes, and mitochondria are found. Microtubules are fairly common and run parallel to the long axis of the cell. As the lens fibers mature, the nucleus and cytoplasmic organelles disappear. In the lenticular nucleus, only the fine dense material remains. The plasma membrane is also disrupted, although the tight junctions persist.

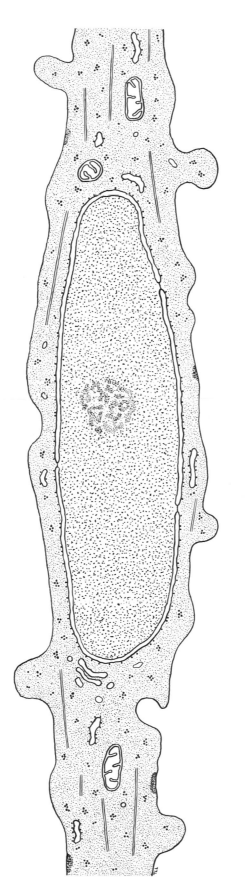

173 — LACRIMAL GLAND CELL

The lacrimal gland is a tubuloalveolar type of gland, producing a serous secretion that moistens and lubricates the surface of the eyeball and washes out foreign particles. The acini of the lacrimal gland are lined by epithelial cells that show the characteristic pattern of secretory cells. The cells are pyramidal in shape with a basal nucleus. Scattered microvilli occur on the apical surface bordering the lumen of the acinus. Some villi and interdigitations occur on the lateral and basal surfaces as well. The rough-surfaced endoplasmic reticulum is extensive, especially in the base of the cell. Ribosomes occur free in the hyaloplasm. The supranuclear Golgi complex is well-developed. Nearby are vacuoles containing flocculent material. The apex of the cell is filled with secretory granules (SG) with a heterogeneous content. Some contain flocculent material, while others are dense or opaque. Mitochondria are elongated and not unusual. Cytoplasmic filaments are present, especially in the apex of the cell. The nucleus contains a nucleolus, and the chromatin is condensed into clumps of heterochromatin. The lacrimal secretion contains protein and lysozyme, an antibacterial enzyme.

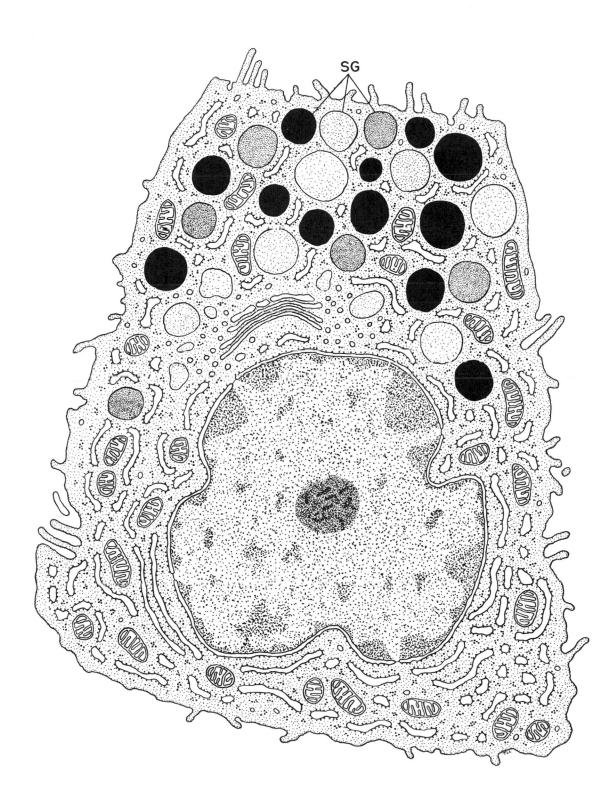

SG

REFERENCES

Bairati, A., Jr. and N. Orzalesi, 1966. The ultrastructure of the epithelium of the ciliary body. A study of the junctional complexes and of the changes associated with the production of plasmoid aqueous humor. Z. Zellforsch., 69:635–658.

Cohen, A. I., 1965. Some electron microscopic observations on interreceptor contacts in the human and macaque retinae. J. Anat., 99:595–610.

Dowling, J. E. and B. B. Boycott, 1966. Organization of the primate retina: electron microscopy. Proc. Roy. Soc. London (Biol.), 166:80–111.

Fine, B. S. and L. E. Zimmerman, 1962. Müller's cells and the "middle limiting membrane" of the human retina. Invest. Ophthal., 1:304–326.

Fine, B. S. and L. E. Zimmerman, 1963. Observations on the rod and cone layer of the human retina. Invest. Ophthal., 2:446–459.

Kühnel, W., 1968. Vergleichende histologische, histochemische und elektronmikroskopische Untersuchungen an Tränendrüsen. III. Schaf. Z. Zellforsch., 87:31–45.

Kuwabara, T., 1968. Microtubules in the lens. Arch. Ophthal., 79:189–195.

Richardson, K. C., 1964. The fine structure of the albino rabbit iris with special reference to the identification of adrenergic and cholinergic nerves and nerve endings in its intrinsic muscles. Amer. J. Anat., 114:173–206.

Tormey, J. McD., 1966. The ciliary epithelium: an attempt to correlate structure and function. Trans. Amer. Acad. Ophthal. Otolaryng., 70:755–766.

EAR

174—ORGAN OF CORTI: HAIR CELL

Two types of hair cells are found in the organ of Corti: the inner hair cells and the outer hair cells. The outer hair cells, illustrated here, are tall, columnar cells with a basal nucleus and a specialized receptor surface. The inner hair cells are basically similar in cytoplasmic structure but are goblet-shaped, with a wide base tapering to a constricted neck region.

Each sensory cell has over 100 sensory hairs (SH) on its surface. The hairs are arranged in parallel rows that form a **W** when viewed from the top. In longitudinal section, the hairs resemble a baseball bat, with a broad distal end tapering to a constricted base. The tips of the tallest hairs are in contact with the tectorial membrane (TM). This structure consists of a finely granular to filamentous outer margin and an inner zone of coarse filaments. Toward their base the sensory hairs have an axial core that penetrates the cuticular plate as a rootlet (Rt) and sometimes passes through it into the cytoplasm. The cuticular plate (CP) is a semicircular mass of finely filamentous material that appears to be a compact terminal web. Its broad side is applied to the surface plasma membrane. A narrow zone of cytoplasm reaches the surface on each side of the cuticle. In one of the cuticle-free zones, a basal body (BB) and a centriole (Ce) are found, but there is no associated cilium in the adult. Coarse aggregates of filamentous material are present along the junctional complexes of the lateral, apical borders.

The apical cytoplasm below the cuticular plate contains a number of organelles. Dense bodies are quite common and some are clearly lysosomes. Their content varies from granular material with droplets to nearly opaque material. A few flattened channels of rough-surfaced endoplasmic reticulum are present. In addition, some smooth lamellae are arranged in a concentric whorl that may correspond to Hensen's body. A Golgi apparatus occurs in the apical region and there are many small vesicles and smooth irregular tubules and channels in the same area.

Below this apical region, organelles are not as densely packed and consist mainly of free ribosomes and vesicular elements. The hyaloplasm has little density to it. The smooth endoplasmic reticulum forms a row of subsurface cisternae (SsC) along the lateral borders. Mitochondria are lined up in a row along the inner sides of the cisternae. The nucleus is located in the lower third of the cell and contains masses of condensed chromatin. Mitochondria are abundant in the basal cytoplasm.

Nerve fibers terminate on the basal surface of the sensory cells. Two types of receptoneural junctions can be distinguished. The first type (1) is relatively simple in structure and is formed by the endings of the afferent cochlear nerve fibers. The endings are small boutons separated from the hair cell by an irregular cleft. Small synaptic vesicles are contained in the terminal. Vesicles and irregular tubules are found in the sensory cell cytoplasm, but subsurface cisternae are lacking. A dense bar, called a synaptic bar (SB), surrounded by vesicles is found in some species opposite this type of ending. Localized condensations of cytoplasm occur symmetrically over short regions beneath the plasma membranes of the nerve fiber and sensory cell. The second type of junction (2) is formed by the nerve terminals of the efferent olivo-cochlear tract. These endings are considerably larger than the first type and are more complex in structure. Large numbers of synaptic vesicles (SV) and a few dense core vesicles are found. Mitochondria are abundant proximally in the fibers. The synaptic cleft is regular in width. In the

(Continued)

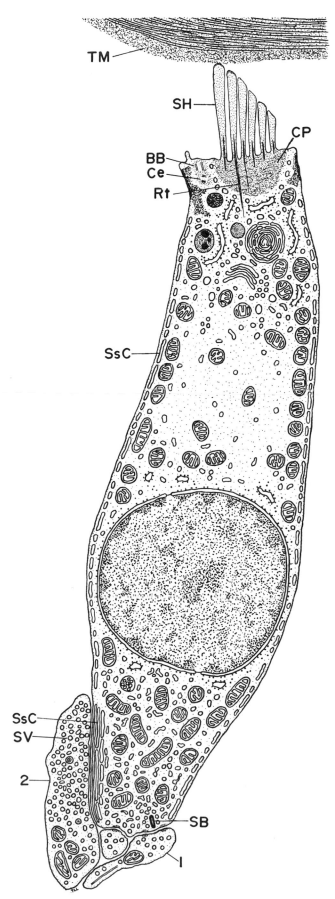

TM

SH

CP

BB
Ce
Rt

SsC

SsC
SV

2

SB

I

174—ORGAN OF CORTI: HAIR CELL (*Continued*)

hair cell cytoplasm, one to three smooth, flattened, cisternae (SsC) are situated just below the surface of the membrane.

Vibrations in the tympanic membrane produced by sound waves are transmitted through the chain of auditory ossicles to the fenestra ovalis and then to the perilymph of the cochlea. Vibrations transmitted to the basilar membrane result in bending of the hairs of the sensory cell in relation to the tectorial membrane. Displacement of the tectorial membrane is accompanied by depolarization of the hair cells. This change in potential may cause a chemical mediator to be released in the basal region of the hair cells, which in turn stimulates the nerve endings. In this manner, according to this theory, the mechanical energy produced by the stimulus is transduced to the electrical energy of the nerve impulse.

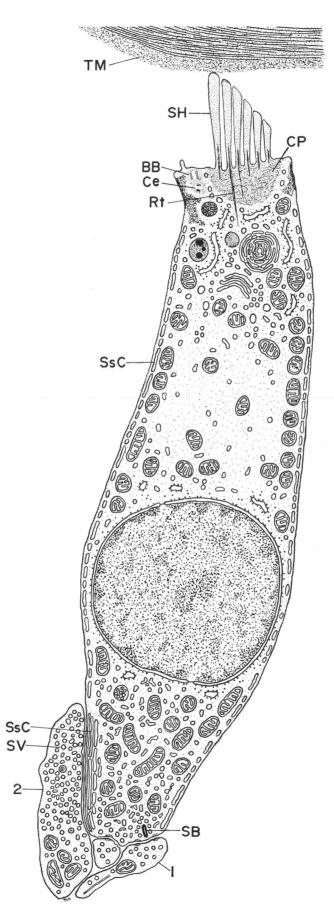

TM

SH

CP

BB
Ce
Rt

SsC

SsC
SV

2

SB

I

175—ORGAN OF CORTI: SUPPORTING CELL

The second component of the organ of Corti is the supporting cells, of which there are several types (inner and outer pillars, inner and outer phalangeal cells, border cells, cells of Hensen, cells of Claudius, and cells of Böttcher). The supporting cells have a number of common cytoplasmic characteristics and differ primarily in their shape and their position relative to other cells. The outer phalangeal cells or cells of Deiters are shown here as an example of this class of cells. These cells act as supporting elements for the outer hair cells. They consist of a body resting on the basement lamina, and a thin stalk extending upward which terminates in a flat process or phalanx. The lower third of the outer hair cell is embedded in the upper portion of Deiters' cell body.

The most conspicuous feature of these cells is a prominent bundle of microtubules (Mt) extending from the base to the terminal phalanx. The microtubules are closely packed in parallel and correspond to the fibrils observed with the light microscope. It seems likely that these structures are responsible for the maintenance of shape of the highly asymmetric supporting cells. Toward the lateral margins of the organ of Corti, where the cells become progressively lower in height, microtubules become less prominent.

The distal phalanx forms contact specializations with adjacent supporting cells and hair cells. A junctional complex with a *zonula occludens* (ZO) and a *zonula adherens* (ZA) is formed between adjacent cells. In addition, there is a prominent accumulation of dense filamentous material in the cytoplasm applied to the lateral surfaces, into which the microtubules project and terminate. A similar accumulation of material was seen in the hair cells and this band below the epithelial surface constitutes the reticular lamina of the organ of Corti.

The nucleus is situated in the upper portion of the cell body. As in the hair cells, the hyaloplasm is of low density and organelles are sparsely distributed. A stack of rough-surfaced cisternae occurs in the supranuclear cytoplasm. Dense bodies or lysosomes are fairly common. Ribosomes, vesicles, smooth membranes, and mitochondria are scattered in the cytoplasm. The Golgi apparatus is not conspicuous.

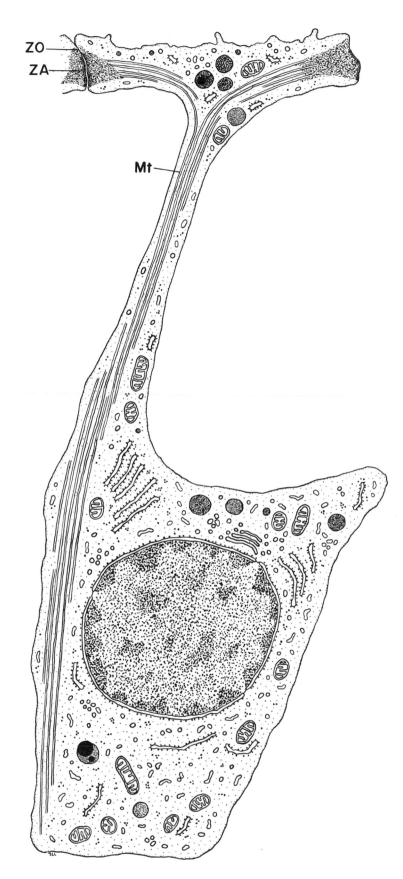

176—STRIA VASCULARIS: MARGINAL CELL

The stria vascularis is a thick, vascularized epithelium covering the outer wall of the cochlear duct. The epithelium is stratified and composed of three cell types: marginal cells, intermediate cells, and basal cells.

Marginal cells are large and columnar and line the luminal surface of the stria. The free surface is convex and bears short microvilli at irregular intervals. The nucleus, which contains a nucleolus, is located in the upper portion of the cell. The portion of the cell below the nucleus is highly compartmentalized by a system of processes and infoldings. The base of the cell rests in concavities in the upper surface of the basal cells.

The supranuclear cytoplasm is crowded with large numbers of vesicles (AV) and small vacuoles. Coated invaginations are seen on the free surface, while coated vesicles lie in the apical cytoplasm. A Golgi complex is found above the nucleus, and lysosomes and multivesicular bodies are associated with it. Lipofuscin pigment bodies and lipid droplets may also occur. The large mitochondria are round or oval in profile and have long, closely packed cristae. Elements of endoplasmic reticulum are not prominent and consist of short, isolated, rough-surfaced cisternae. A few free ribosomes and filaments are found in the cytoplasm.

The basal cytoplasm is divided into a complex system of compartments by deep infoldings of the lateral and basal plasma membranes. The compartments are similar to those seen in cells that are believed to be involved in the active transport of fluids and ions, such as the distal convoluted tubule cells of the kidney (Plate 101) and the striated ducts cells of the salivary gland (Plate 68). The endolymph contains an unusually high concentration of potassium, and a low concentration of sodium. The marginal cells, with their large surface area and their mitochondrial energy supply, may secrete potassium ions into the endolymph. The coated pits and vesicles on the surface, on the other hand, could be involved in selective absorption of materials from the endolymph (see also Plates 99, 111, 131).

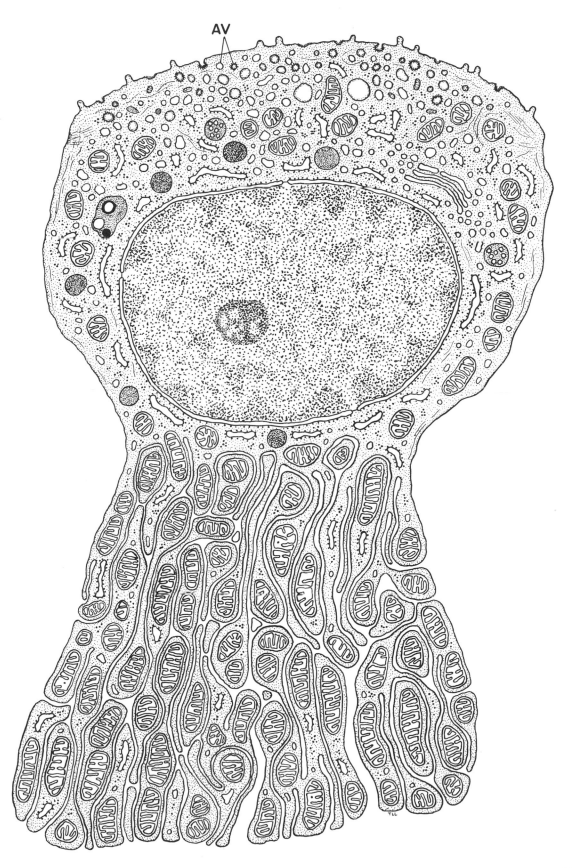

177—STRIA VASCULARIS: INTERMEDIATE CELL

The intermediate cells of the stria vascularis have a considerably different appearance from that of the marginal cells. They are found below the level of the nuclei of the marginal cells and do not appear to reach the luminal surface. They are highly irregular in shape, with branching cytoplasmic processes. The processes extend between marginal cells and may interdigitate with the basal processes of the marginal cells. They also terminate adjacent to capillaries and basal cells. The nucleus, with its nucleolus, is centrally placed. The hyaloplasm has little density. A Golgi apparatus is seen in the juxtanuclear region. Short cisternae of rough-surfaced endoplasmic reticulum and a few free ribosomes are found in the cytoplasm. Smooth-surfaced elements occur as well, and in the processes these structures are sometimes arranged into concentric lamellar arrays (lower right of Plate). Mitochondria are small and rod-shaped with transverse cristae and dense matrix. Lysosomes are fairly common and there is a system of cytoplasmic filaments.

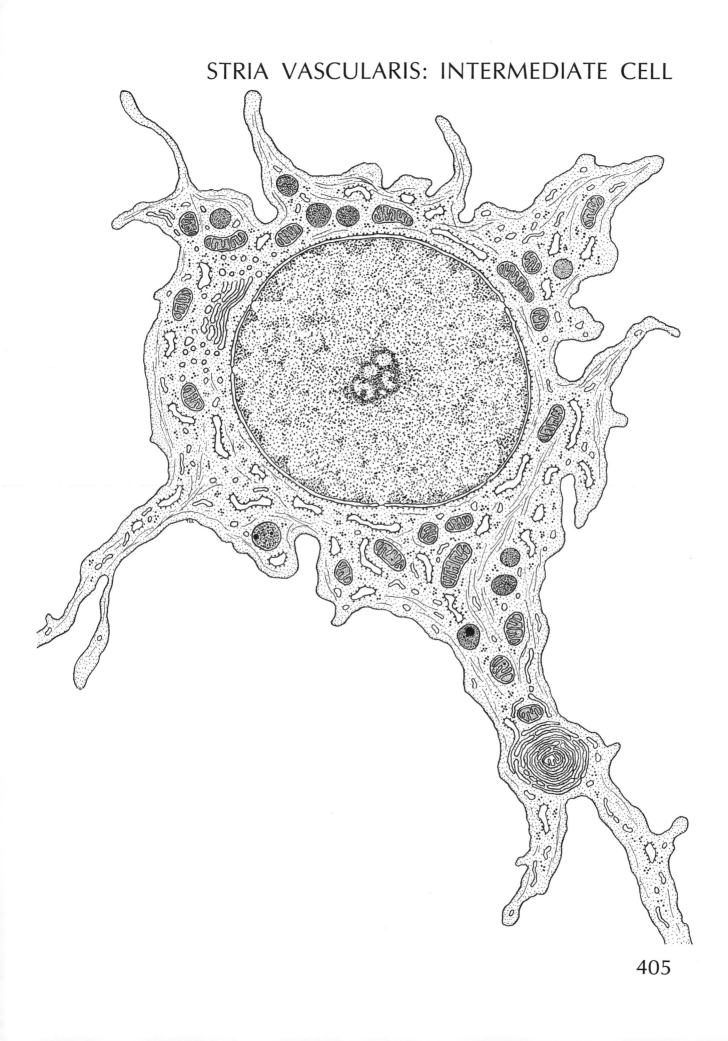

178—STRIA VASCULARIS: BASAL CELL

The basal cells are somewhat flattened and form the bottom layer of the strial epithelium. They do not appear to be separated by a basement lamina from the underlying cells of the spiral ligament. The basal cells are irregular in shape with long cytoplasmic processes. Lateral processes overlap or abut on processes of adjacent basal cells. Apical processes extend between marginal cells and seem to envelop them in a cup-like fashion. Nuclei are centrally located. In cytoplasmic structure, the basal cells are similar to intermediate cells. A Golgi apparatus occurs near the nucleus and elements of endoplasmic reticulum are sparsely distributed. Free ribosomes, mitochondria, and filaments are present. In addition, there are a number of dense bodies, most probably lipoidal or lysosomal in nature. The exact function of both intermediate and basal cells is uncertain. Attention has been called to their similarity to neuroglia in the central nervous system. On the other hand, they also seem comparable to the small basal cells found at the bottom of other epithelia.

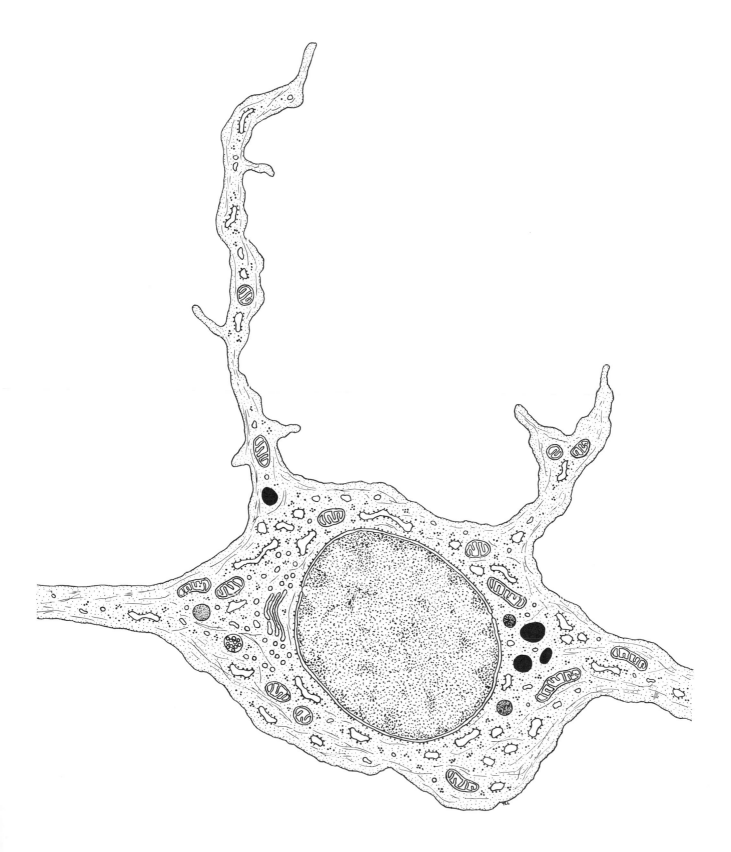

179—VESTIBULAR LABYRINTH: HAIR CELL—TYPE I

The components of the vestibular labyrinth of the internal ear include the utricle, the saccule, three semicircular canals and their ampullae, and the endolymphatic duct and sac. Sensory epithelium is found in the maculae of the utricle and saccule and in the cristae ampullares. It consists of two basic cell types: the supporting cells and the hair cells; and among the hair cells, two types can be further distinguished.

Type I hair cells are flask- or goblet-shaped, with a wide base, constricted neck, and widened apex. The surface of the hair cell bears up to 100 sensory hairs (SH), which are called stereocilia in these cells. They are narrow at their bases and widened at the top, and they appear to represent modified microvilli. The hairs have dense rootlets in their bases which pierce the cuticular plate (CP). In addition to the hairs, a cilium called a kinocilium (Kc) is also found on the sensory surface. It arises from a basal body (BB) situated in a cuticle-free zone of apical cytoplasm. The longest sensory hairs are situated adjacent to the kinocilium. The cilium and long hairs extend into the otolithic membrane of the utricle and saccule, or into the cupulae of the cristae ampullares. The cupulae are gelatinous masses with a fibrillar appearance. The otolithic membrane contains crystals of calcium carbonate or otoliths (Ot). It is thought that gravitational pull on the otolithic membranes causes movement of the sensory hairs, which in turn results in a change in polarization of the cell.

The round nucleus is located in the dilated basal region of the cell. Mitochondria are abundant in the basal regions. A Golgi apparatus occurs near the nucleus. Elements of rough-surfaced endoplasmic reticulum are few in number, but vesicles and smooth cisternae are common. Lysosomes are also present in the sensory cell cytoplasm. Microtubules (Mt) are numerous and in the neck run upward into the apical region. A mass of finely granular to filamentous material forms the cuticular plate (CP) in the apex of the cell. In addition, accumulations of coarser, denser material are applied to the lateral membranes adjacent to the junctional complexes.

The two types of sensory cells are distinguished primarily on the basis of the configuration of the nerve fibers terminating upon them. On type I cells, the vestibular nerves end as chalices that enclose all of the cell except the apex in a cup-like fashion. The synaptic gap is most uniform at the base of the cell although in places the membranes come into close apposition. The gap is more irregular toward the apex of the cell. Mitochondria are common in the axoplasm, but vesicles are relatively few in number.

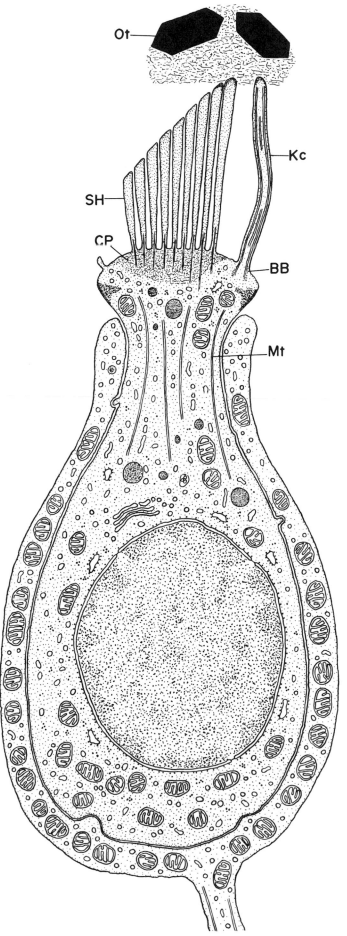

Ot

Kc

SH

CP

BB

Mt

409

180 — VESTIBULAR LABYRINTH: HAIR CELL — TYPE II

Type II hair cells are columnar and basically similar in cytoplasmic structure to the type I cells. The Golgi apparatus is better developed and there are a greater number of small vesicles and multivesicular bodies. Microtubules are not as prominent. As mentioned, the cells differ primarily in the nature of the nerve endings, and in type II cells the endings are bouton-type terminals. Two kinds of terminals can be distinguished. The first (1) contains mitochondria, vesicles, and some larger dense core vesicles. There are some slight, localized thickenings of the plasma membranes of the nerve and sensory cell. In the cytoplasm of the sensory cell, a synaptic bar (SB) commonly occurs. This structure is a dense rod with vesicles aligned along its sides. These endings are thought to be afferent. The second type of terminal (2), considered efferent, contains a dense accumulation of synaptic vesicles (SV). Synaptic rods do not occur in the cytoplasm opposite these endings; instead, there is a subsurface cisterna (SsC). The structure of these two endings is quite similar to that of terminals on the hair cells in the organ of Corti (Plate 174).

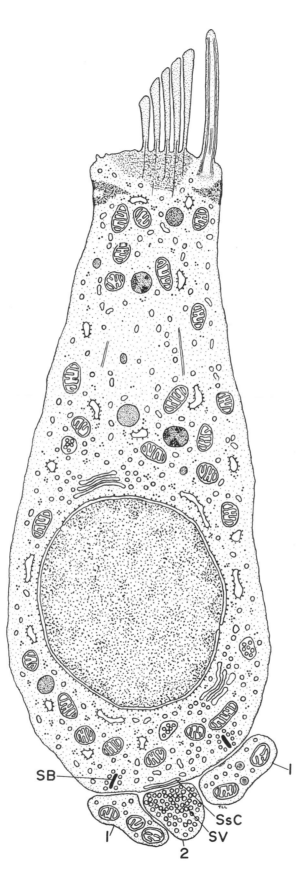

181 — VESTIBULAR LABYRINTH: SUPPORTING CELL

The supporting cells are tall irregular cells interposed among and surrounding the sensory cells. They rest on the epithelial basement lamina and reach the surface. The surface membrane bears a number of small microvilli and a single kinocilium (Kc). Unlike the cilia of the sensory cells, the kinocilium of supporting cells appears to lack the two central fibrils. In addition, a striated rootlet (Rt) extends downward from the basal body. A centriole (Ce) is situated near the basal body. An extensive filamentous terminal web area is found in the apex and constitutes the reticular lamina of the sensory epithelium. Junctional complexes are formed along the apical lateral borders.

The nucleus is located at the base of the cell. A prominent band of microtubules (Mt) runs from the base to the terminal web area. The rest of the cytoplasm has a sparse population of organelles (cf. Plate 175). A small Golgi apparatus is found near the nucleus, and there are scattered short cisternae of endoplasmic reticulum, ribosomes, vesicles, and mitochondria. Lysosomes are present as well as some smaller dense granules that are possibly secretory in nature.

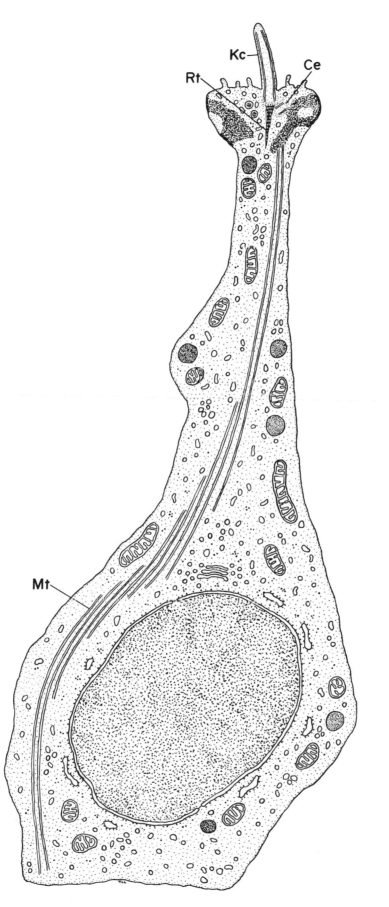

182—VESTIBULAR LABYRINTH: DARK CELL OF THE CRISTA AMPULLARIS

Specialized cells are found in the lining of the ampullae near the bases of the cristae. They are variable in shape and characterized by elaborate folding of the basal plasma membrane. Large mitochondria fill much of the cytoplasm of the basal processes. The nucleus is located toward the middle of the cell and the supranuclear cytoplasm is uncomplicated. A small Golgi apparatus is present along with a pair of centrioles. The endoplasmic reticulum consists of short, rough-surfaced cisternae. A number of lysosomes are found and mitochondria and ribosomes are scattered in the cytoplasm. The apical surface is relatively smooth with only an occasional short microvillus. Vesicles are found in the apical cytoplasm and may be continuous with the surface membrane. Dilated vacuoles (Vac) are also found and have been interpreted as evidence of secretory activity. A similarity has been noted between these vestibular dark cells and the marginal cells of the stria vascularis (Plate 176). Both have complicated infoldings and apical vesicles. Like the marginal cells, the dark cells may have a role in the production of endolymph.

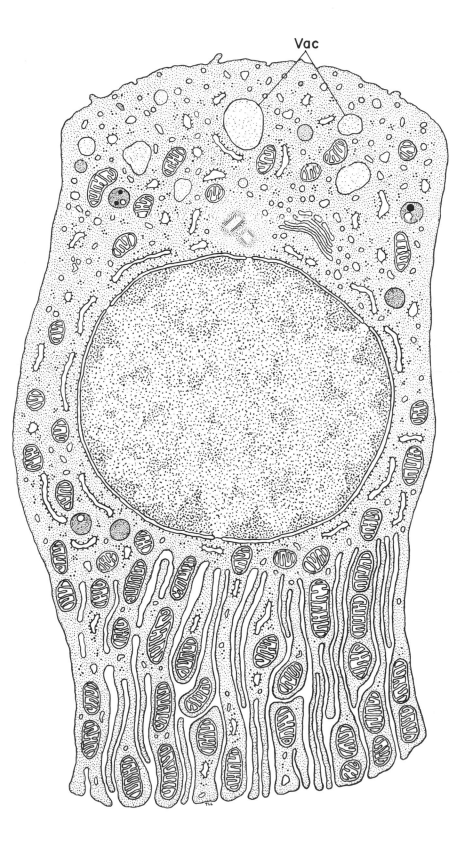

Vac

183—VESTIBULAR LABYRINTH, ENDOLYMPHATIC SAC: LIGHT CELL

The endolymphatic sac is located on the surface of the petrous part of the temporal bone and is connected by a duct to the membranous labyrinth at the junction of the saccular and utricular ducts. Two cell types can be distinguished in the epithelial lining of the endolymphatic sac: the light cells and the dark cells.

The light cells are cuboidal to columnar in shape with a round, central nucleus containing a nucleolus. Microvilli extend from the apical surface and small invaginations of the plasma membrane occur at their bases. Vesicles are also common in the apical cytoplasm. A typical junctional complex is formed between neighboring cells. A number of processes or villi are seen along the lower portions of the lateral borders. Infoldings also occur in some places along the basal surface. A Golgi apparatus with associated vesicles and vacuoles occurs in the cytoplasm above the nucleus. The endoplasmic reticulum is poorly developed, consisting of short, isolated, ribosome-studded cisterns. Free ribosomes, however, are more abundant and tend to occur in clusters. Mitochondria are round to oval in shape with transverse cristae and a dense matrix. A number of lysosomes occur in the cytoplasm. The light cell is thought to be involved in the reabsorption of endolymph.

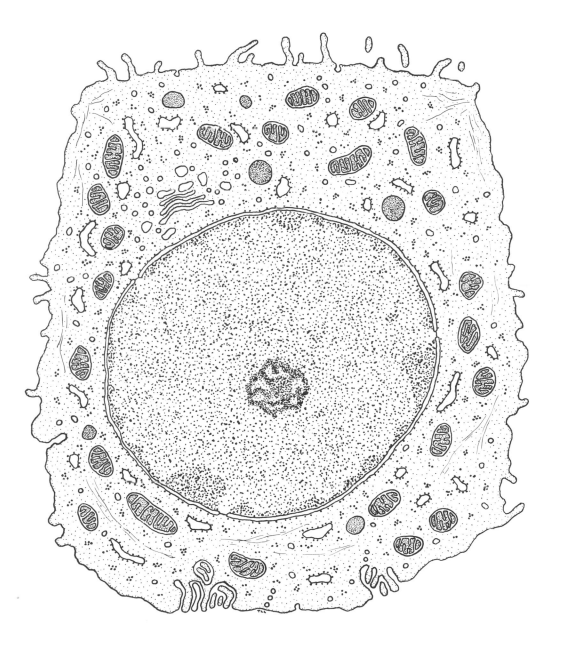

184—VESTIBULAR LABYRINTH, ENDOLYMPHATIC SAC: DARK CELL

Dark cells have a denser hyaloplasm than do light cells and tend to be pyramidal in shape with a narrow base. The basal surface is somewhat irregular with a number of processes or infoldings. The nucleus is located nearer the apex of the cell and is oval with several indentations. The chromatin shows a greater degree of condensation than that of the light cell. A few microvilli occur at the surface, but there are not nearly as many as in light cells. A Golgi apparatus is located in the upper portion of the cell. There are some flattened cisternae of rough-surfaced endoplasmic reticulum and some free ribosomes. Mitochondria and a few lysosomes are distributed in the cytoplasm. In the dark cell, there is a system of cytoplasmic filaments running between organelles. Dark cells have been shown to phagocytose particulate materials within the endolymphatic sac and are thought to constitute a defensive mechanism for the inner ear.

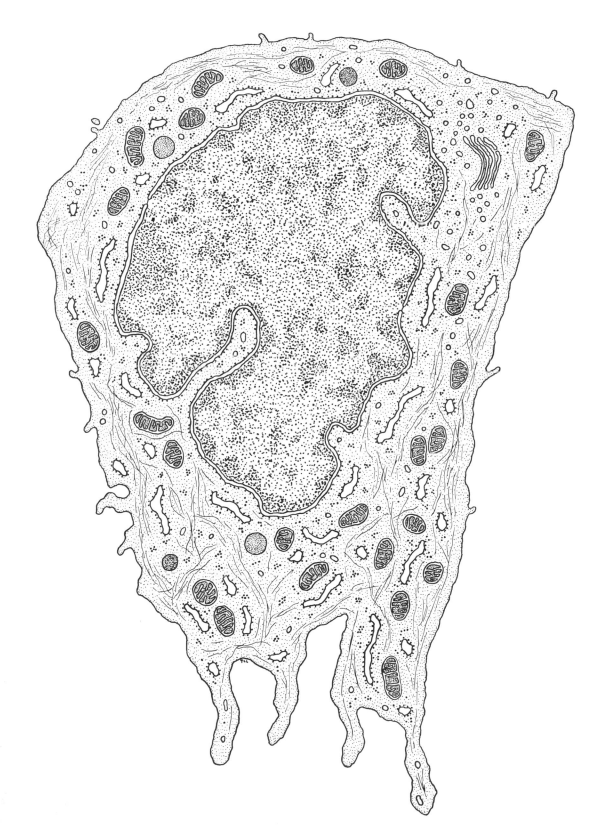

INDEX

423

INDEX

INDEX

427

INDEX

INDEX

INDEX

433

INDEX